ISBN: 978-1-7638817-0-9

Book Cover by Bojan R

1st. edition 2025

To my children Robert, Naomi, Melinda and Julia, with love

Contents

How to Prosper in an AI World

A Non-Technical Guide to Thriving in the Age of Artificial Intelligence

Otto Kern

Aligned Intelligence

Preface

Writing a book requires a certain audacity, a bold conviction that one's thoughts matter, and the hope that, amidst countless competing alternatives, someone might choose to read it. Publishing is an act of courage, an acknowledgment of its own presumptions, yet a step forward nonetheless. So why take this leap? Why add another voice to an already overflowing conversation? Because the times we're living in demand it!

We are poised to witness a technological revolution unprecedented in human history. Artificial Intelligence has grown far beyond a trendy buzzword; it is now everywhere. Quietly yet profoundly, it is reshaping industries and economies, weaving itself into the fabric of everyday life, often unnoticed. But while AI has the potential to enrich our lives, it also risks narrowing our perspectives. Algorithms filter what we see and hear, trapping us in cozy bubbles that affirm our existing beliefs rather than challenging us to think more broadly.

This book is an attempt to break through those barriers, not just to provide information but to offer a balanced perspective on AI: its promises, its risks, and the profound ways it is altering our world.

Another reason for writing this book stems from a question I often hear: "Do I need formal education to understand AI?" The simple answer is no, but that's not the whole story. While the wealth of free resources makes AI seem accessible, that same abundance creates its own challenges. How do you separate what is valuable from what is noise? How do you make sense of a field so vast and fast-evolving, especially when you may not yet know what you don't know?

You may not need a degree to grasp the essentials of AI, but you do need structure, a clear guide to help navigate its complexities. My aim is to provide an approachable roadmap that introduces the foundations of AI and points you toward areas you may wish to explore in greater depth.

This book serves several purposes. It offers a high-level overview of AI's current landscape, introduces key technical concepts in accessible terms, and explores some of the pressing philosophical questions about AI's place in society. Above all, it provides practical insights—tools you can use not just to navigate but to thrive in a world increasingly shaped by AI. Whether you are a professional wondering about AI's impact on your future, an entrepreneur seeking new opportunities, or simply someone who feels overwhelmed by the pace of change, this book is for you.

What makes this book particularly special to me is how it was created. From brainstorming ideas to refining concepts, researching details, and even formatting the pages, AI has been my collaborator at every step of the way. This partnership embodies the very essence of what this book advocates: the transformative potential that emerges when humans and AI combine their respective strengths to create something greater than either could achieve alone.

This book is a testament to that partnership and to the strong conviction that when technology is used mindfully, it empowers rather than replaces us.

As you turn these pages, I hope you discover not just information, but clarity, inspiration, and a renewed sense of agency. This book is about thriving in an AI-driven future, not merely adapting to it. Writing it has been a journey of growth and discovery, and I'm excited to share this adventure with you. Let's navigate the possibilities of AI together with curiosity and confidence.

Otto Kern,

(2025, West Perth, Australia)

Acknowledgements

I want to express my deep gratitude and appreciation to the countless courageous voices, both within and beyond the halls of academia, who share their knowledge and insights despite the pressures they often face. These individuals, many working quietly with modest audiences, create valuable content fueled by their passion for progress and their genuine concern for our collective future. Though too numerous to name individually, their influence has been profound, both on this book and on the world at large.

Their unwavering commitment to sharing ideas and expertise inspired me to add my own voice to this important conversation. They have shown me that significant advances in human understanding need not always emerge from conventional or established sources. Genuine concern, dedication, and a willingness to share can create ripples that inspire others to think, question, and grow.

This book is shaped by their example and embodies the same spirit of intellectual generosity that I believe will help us all thrive in an AI-driven future.

Lastly, and most importantly, I want to acknowledge the unwavering support of my wife, Natalia. Thank you for your selfless encouragement during the challenging times of the past eight years, as I transitioned from the business world to embark on deep philosophical exploration and academic pursuits—a journey that often left me distant, absent, and disengaged. Your patience and belief in me have been a constant source of strength and motivation.

To all those who have shared their wisdom and passion, thank you. You have made this journey possible.

-Author

Introduction

A World on the Verge of Transformation

> "Any sufficiently advanced technology is indistinguishable from magic."
>
> Arthur C. Clarke

It was March 31, 1880. As dusk settled over Wabash, Indiana, the townsfolk gathered in the square, their breath clouding in the crisp evening air, hearts pounding with anticipation. Rumors had spread, talk of a light so powerful it could banish the dark and turn night into day. Skepticism lingered; after all, tales were one thing, witnessing quite another.

When the four arc lamps lit up, filling the area with a brilliant, almost heavenly glow, the audience gasped. Some stared in wonder, others stepped back in fear. For most, it felt like witnessing a miracle.

But miracles don't come easy. What the people of Wabash couldn't see that night was the years of relentless effort that had led to this moment. Behind the dazzling display lay a history of failures, rejection, and unwavering determination. Decades earlier, Benjamin Franklin's kite experiment hinted at electricity's potential but left the world wondering how it could be harnessed. Then came visionaries like Michael Faraday and Joseph Henry, quietly solving puzzles others considered unsolvable; their groundbreaking work, often in obscurity, was dismissed as useless or unattainable.

Charles F. Brush, the man behind the arc lighting system that illuminated Wabash, faced his own share of hurdles. His early prototypes sputtered and flickered, and the weak, uneven lighting failed to inspire trust. Investors ridiculed his ambition, labeling it unrealistic or outright dangerous. Nonetheless, he and his colleagues persevered, frequently laboring by candlelight to create a brighter future, literally.

When those arc lamps eventually lit up the Wabash square, they represented far more than technological achievement. They served as beacons of human perseverance, creativity, and the refusal to let failure have the final word. That night, as the people of Wabash looked on, some marveled at the brilliance, while others worried about what it might mean for their livelihoods, their beliefs, and their way of life. It was their first glimpse of a future beyond their full comprehension—one that was dazzling, overwhelming, yet utterly inescapable.

Today, that light has expanded into every corner of the planet. Electricity, once a groundbreaking marvel, has become so enmeshed in our lives that it feels invisible. It powers our homes, businesses, and hospitals. It ensures the smooth operation of commerce and industry. It connects continents, enabling instantaneous communication that was once the realm of science fiction. It is the subtle force that drives every video we watch, every email we write, and every algorithm that makes sense of our digital universe. And yet, for all its ubiquities, electricity is still a kind of magic. Its absence, whether in a power outage or in imagining a world without it, reminds us of just how vital it is. Like the people of Wabash, we may no longer marvel at the light itself, yet its impact on our lives remains profound, subtly shaping everything we do, often without our notice.

I wasn't there in 1880 to witness the electrification of Wabash, but I can empathize with the awe and uncertainty its residents must have felt. Strangely enough, I felt something similar when I first encountered artificial intelligence. It wasn't during its earliest days, but later, when AI began to evolve from an abstract concept to a force with tangible potential.

My own journey with AI began in the late 1980s or early 1990s in Brisbane, Australia. Back then, there was no Google to consult. So, when I first heard the term "Artificial Intelligence," I turned to the only university-educated person I knew, our local preacher. "What exactly is artificial intelligence?" I asked him with genuine curiosity. "What's it all about?" He looked at me with a bemused expression and said, "AI? It's pseudo-science. A cult. Just a bunch of lunatics trying to play God by attempting to recreate humans." At the time, his perspective seemed reasonable, even authoritative. It shaped how I viewed AI for years. Whenever I heard the term, I disregarded it as some wild, impractical endeavor, a collection of dreamers pursuing an unachievable, possibly hazardous concept. But just as the people of Wabash eventually came to see the light not as a threat but as a transformative force, my understanding of AI has evolved. This book is about that evolution, both personal and societal, and how AI, much like electricity, is becoming an inescapable part of our world, one that challenges us, empowers us, and invites us to see the future in a new light.

Fast forward to November 2022, and the phrase "Artificial Intelligence" resurfaced for me in a completely different context. My social media feed contained a brief footage of someone talking with a conversational bot known as ChatGPT. What caught my attention wasn't just the bot's wit but the way it responded: fluid, natural, and, in a way that felt unsettling, deeply human. I was utterly astonished. I had to experience it myself. I don't remember my precise first prompt, but I'll never forget how I felt when I received the response. I felt a flood of feelings all at once: excitement, bewilderment, curiosity, and, yes, terror. I sat there staring at my laptop, trying to absorb what had just happened. It wasn't just a brilliant tool or a scripted interaction. For the first time, I felt the power of something truly transformative, an unstoppable force unleashed, much beyond what I could have anticipated. The pastor's words from decades earlier echoed in my mind: "Lunatics playing God by trying to recreate humans." Were they now well on their way to succeeding?

What struck me wasn't just the technological brilliance but the unmistakable sense of a watershed moment. That single interaction set something in motion. Over the next few days, I found myself reassessing my priorities and reconsidering the course of my life. By the end of the week, I'd made the bold decision to change course entirely. I left behind what I was doing, enrolled in a university program to learn the fundamentals of AI, eventually pursuing a master's degree. My current study focuses on the commercial applications of this groundbreaking technology. Artificial intelligence, like electricity in its early days, has evolved from an intriguing curiosity to a multi-purpose technology. And much as electricity transformed every aspect of life, AI is beginning to weave itself into the very fabric of our existence.

My first interaction with ChatGPT made me realize something profound: AI is the new electricity. It is a force that will transform the way we live, work, and connect. AI, like electricity, is unaffected by our thoughts about it. It is neither intrinsically good nor evil. It's a transforming force waiting for us to determine how to use it. History shows that the true potential of disruptive technologies often goes far beyond our initial intentions and expectations. Who could have predicted, when electricity first lit a few arc lamps, that it would lead to smartphones, life-saving medical devices, and global connectivity? Similarly, we still don't fully understand the ramifications of AI. What we do know is that AI holds the promise to enhance creativity, automate the mundane, and tackle challenges we once considered insurmountable. But as with any powerful tool, its ultimate impact depends on the choices we make. Will we wield it responsibly? Will we mold it in ways that benefit humanity as a whole?

This book is not about observing the AI revolution from the sidelines. It's a call to action, a guide for stepping into the arena with curiosity and intention. By understanding AI's strengths, limitations, and ethical dimensions, we can move from passive spectators to active participants in shaping its trajectory. The decisions we make now won't just shape the future of technology; they'll influence the way we work, learn, connect, and live together as a society. AI is not waiting for us to catch up; it's already here, evolving rapidly. Remaining on the sidelines means risking irrelevance. However, by interacting with it, learning, adapting, and contributing, we position ourselves as co-creators of a future in which AI augments rather than replaces human capability.

This book is structured to guide you through the fascinating world of AI, starting with its history and foundational concepts, moving through its current applications and societal implications, and concluding with practical strategies for thriving in an AI-driven future. Along the way, we'll explore the transformative potential of AI, its ethical dilemmas, and its role in shaping industries, governments, and personal lives. Each chapter builds on the last but can also stand alone, allowing you to navigate the material in a way that best suits your interests and needs. With this in mind, let us begin this journey together.

Chapter 1

Foundations of Understanding AI

Unpacking it's Intricacies

"What we call the present is given shape by an accumulation of the past."

Haruki Murakami

Step into any coffee shop today, and you'll notice a scene that would have seemed like pure science fiction just a decade ago. Someone is deep in conversation with ChatGPT, another is fine-tuning AI-generated art, and a student is getting homework tips from an AI tutor. What once belonged to the realm of futuristic fantasy landscapes has become our everyday reality. But how did we get here? And more importantly, what does this transformation mean for our understanding of human nature, our values, and society as a whole?

We'll take an incredible voyage through the development of artificial intelligence in this chapter, not only as a technological phenomenon but also as a force that has continuously challenged our self-perception. Every AI development has confronted us to reconsider what it means to be human and to reevaluate beliefs we once took for granted.

The Early Days: When Machines Began to "Think"

In the mid-20th century, visionaries like Alan Turing and John von Neumann asked a provocative question: Could machines think? At the time, the idea seemed both audacious and borderline heretical. Logical reasoning was seen as a hallmark of humanity, inseparable from consciousness itself. Yet when early computers began performing logical operations, they weren't just solving math problems; they were opening a door to something far deeper. If machines could reason, what other qualities, once thought uniquely human, might they possess?

From Rules to Expertise: The Rise of Expert Systems

Fast-forward to the 1970s and 1980s, when expert systems emerged. These were programs capable of solving complex problems in specific fields, often matching or even outperforming human specialists. Suddenly, machines weren't just calculating; they were diagnosing diseases, analyzing legal documents, and identifying chemical compounds. This was a philosophical provocation as much as a technical accomplishment. If machines could replicate expertise acquired through years of study and practice, was human expertise no more than a sophisticated application of rules?

The Strategic Mind Faces Its Match

The chessboard became the stage for another groundbreaking moment in 1997 when IBM's Deep Blue defeated world champion Garry Kasparov. People have long viewed chess as the pinnacle of human strategic thinking, embodying creativity and intuition. When a machine triumphed in this domain, it wasn't just a loss for humanity's best player; it was a challenge to the very nature of human intelligence. Could creativity and intuition be nothing more than complex computational processes?

Learning to Learn: The Machine Learning Revolution

The 2000s brought us machine learning, a game-changer in the world of AI. Unlike earlier systems that relied on explicit programming, these algorithms could learn from experience, improving with each iteration. Machines were now mirroring an ability, once considered a hallmark of biological intelligence. It forced us to confront a startling possibility: perhaps learning isn't as uniquely human, or even biological, as we once thought.

Machines That "See"

When computer vision systems began outperforming humans in specific tasks, another foundational belief was shaken. Machines could "see" without eyes and interpret visual information without a brain. This achievement raised profound questions: if perception doesn't require a biological body, what does that say about how humans understand the world? The success of these systems suggested that some of our most cherished assumptions about human experience might need rethinking.

Speaking Our Language: The Dawn of AI Communication

Then came the advent of large language models like GPT, capable of generating poetry, stories, and even music while engaging in nuanced conversations. Language has always been humanity's primary tool for expressing thoughts, emotions, and creativity. When machines started wielding this tool with fluency, it brought us face-to-face with unsettling questions: Where does creativity come from? What does it mean to truly understand something? And can machines ever grasp meaning in the way humans do?

The Integration of Intelligence

Today, multimodal AI systems combine text, images, and sound with remarkable ease, creating cohesive representations of the world. These systems mirror how humans integrate sensory information, bringing us closer to a reality where machines interpret the world in profoundly human-like ways. The implications are both thrilling and terrifying.

Emergence: When Machines Surprise Us

Imagine observing an AI system carry out operations or find solutions to issues in ways that its designers never specifically intended. This is precisely what we're seeing with today's large AI models. Their sheer scale and complexity give rise to unexpected behaviors, capabilities that seem almost like self-reflection or creative problem-solving. These moments force us to pause and ask profound questions: Could these machines be nudging us closer to understanding consciousness and cognition in ways we hadn't anticipated?

Looking Forward: Rethinking What It Means to Be Human

We find ourselves at a pivotal moment in history. Many abilities we once considered uniquely human, like reasoning, creating, solving problems, are now being imitated, and in some cases, rivaled by machines. This realization challenges us to dig deeper. If these traits no longer define us, what truly makes us human? What sets us apart?

While this book won't claim to answer such sweeping philosophical questions, it's hard to imagine truly thriving in the age of AI without some reflection on these themes. With that in mind, let's explore some foundational ideas about artificial intelligence, starting with its very name: "Artificial Intelligence."

The Birth of AI: Artificial and Intelligent?

The term "Artificial Intelligence" was officially coined in 1956 at the Dartmouth Conference, where pioneers like John McCarthy, Marvin Minsky, Nathaniel Rochester, and Claude Shannon gathered to propose that machines could simulate human intelligence. They envisioned an area of research devoted to creating systems with the ability to think, learn, and solve problems, qualities that have long been linked to human cognition.

The term "artificial" underscores the deliberate design and engineering of this intelligence, which is a result of human ingenuity rather than biological evolution. But here's where it gets interesting: does labeling AI as "artificial" create an unnecessary divide?

Rethinking "Artificial": A Natural Continuation

The word artificial suggests a separation between what is created by nature and what is crafted by humans, as if nature and humanity were two distinct forces. Yet, if we're part of nature, doesn't everything we create also belong to nature? Consider this: at what point does a naturally occurring material like silicon, extracted from the Earth, become "artificial"? Where exactly do we draw the line?

It's more productive, and perhaps more accurate, to see AI as a natural extension of human creativity, another, albeit new, way that nature expresses itself through us. When we embrace this perspective, AI ceases to be a foreign, alien concept and instead becomes something we can collaborate with, a partner in the ongoing evolution of intelligence.

Intelligence: A Concept Under Construction

Defining intelligence has always been elusive. Most commonly, in its simplest form, we think of it as the ability to learn, adapt, and navigate a given environment successfully. But this definition quickly raises more questions: Does an insect's ability to adapt to its environment make it intelligent? What about a plant that exhibits amazing reactions to stress, light, and gravity? Could a galaxy's capacity to self-organize in accordance with physical principles even qualify it as intelligent?

Our human-centered conception of intelligence is called into question by these instances. They propose that intelligence might be a universal characteristic of complex systems rather than something exclusive to biological beings. Intelligence, in this broader sense, might manifest wherever there's interaction, adaptation, and complexity. In other words, everywhere.

The Spectrum of Intelligence

A useful framework for reconsidering intelligence is provided by the Theory of Embedded Intelligence (TEI). Instead of seeing it as a binary trait, something you either have or don't, it describes intelligence as a spectrum of abilities: sensing, processing, communicating, and acting (SPCA). These abilities are evident across all levels of complexity, from cellular processes to cosmic phenomena.

By this definition, intelligence isn't confined to humans or even living beings. Similar to how gravity subtly shapes the universe, it exists both latently ("implicate" as unrealized potential) and obviously ("explicate" via observable acts). This broader perspective invites us to see intelligence as a fundamental property of reality, one that AI is now forcing us to explore in entirely new ways.

This broader perspective on intelligence and AI does more than simply change how we think about machines. It challenges us to rethink our place in the universe and how we relate to the tools we create. As we move forward, the question isn't just what can AI do? But how can we work with AI to build a future where both thrive together?

An Adaptable Framework for Intelligence

The goal of this discussion is not to pin down intelligence to a single definition but to propose a fluid and adaptable framework. When we treat concepts as instruments rather than immutable truths, we open ourselves to a wider perspective. This mindset allows us to connect seemingly disparate manifestations of intelligence, human, artificial, and beyond, into a unified, cohesive whole. This broader perspective on intelligence and AI does more than simply change how we think about understanding. It equips us to navigate the complexities of our evolving world, appreciate the interconnectedness of existence, and embrace emerging intelligences like AI as part of a broader natural progression.

Perhaps being human is less about what we can do and more about how we exist: how we interpret our conscious experiences, cultivate empathy, act with moral agency, take responsibility, and, most uniquely, derive meaning from our lives. As AI continues to evolve, exploring these deeper dimensions of human nature shifts from a philosophical curiosity to a practical necessity.

The instruments we create reflect our values, aspirations, and quest for progress. But they do more than mirror us; they shape us in return, subtly influencing how we think, act, and perceive the world in often unexpected ways. AI is a prime example of this dynamic. It's a tool born from our ambition to replicate and enhance intelligence, yet it challenges some of our most fundamental beliefs about ourselves and our place in the world.

Understanding the AI Revolution: Transforming Sectors and Society

In 2016, AlphaGo, an AI system developed by DeepMind, stunned the world by defeating the world's best Go player. Go was long regarded as the ultimate test of human strategy and intuition because of its almost endless complexity. However, AlphaGo's victory reshaped our understanding of creativity, elevating AI's role beyond games. Its novel tactics, which were praised by experts as "beautiful" and "inspirational," showed how AI might go beyond strict problem-solving and offer creative answers that even professionals couldn't foresee.

This incident brought to light a more general reality: AI expands and redefines current frameworks, both physical and mental, rather than just fitting into them. It is a disruptive force that is changing society, industry, and our perception of what is feasible. Unlike earlier technologies like the steam engine or telephone, which addressed specific industries, AI adapts to countless contexts, embedding itself into the very fabric of daily life and revolutionizing nearly every sector.

The Adaptability and Learning Potential of AI

AI's ability to learn and adapt sets it apart from traditional tools that follow fixed, pre-defined rules. Systems like AlphaGo not only master tasks but also evolve dynamically, tackling unforeseen and previously unseen challenges with ingenuity. This adaptability positions AI as a collaborator, not just a tool, capable of innovating alongside humanity.

Consider healthcare; AI systems such as IBM Watson Health analyze vast medical datasets to assist in diagnosing conditions and recommending treatments. Radiology tools like Zebra Medical Vision detect early signs of diseases with unparalleled speed and accuracy, while AI-powered platforms enhance pandemic preparedness by modeling virus spread and aiding real-time decision-making. Moreover, AI's contributions extend to genomics, where technologies like Illumina accelerate the identification of genetic markers, unlocking new possibilities in personalized medicine.

Beyond healthcare, AI transforms industries like agriculture, where it optimizes irrigation, monitors crop health, and reduces waste. In finance, algorithmic trading platforms analyze market trends in real time, making data-driven decisions to optimize investments. In logistics, AI predicts demand and streamlines just-in-time supply chains (JIT), minimizing inefficiencies. Even creative fields are benefiting from tools like Adobe Sensei, which amplify artistic potential by merging human creativity with technological precision.

Rethinking Our Relationship with Technology

AI challenges long-held assumptions about technology's role in our lives. Historically, machines extended human abilities as passive tools and instruments under direct control. AI, however, learns, adapts, and initiates actions, evolving from a mere instrument to a collaborative partner. This shift requires us to reconsider trust, transparency, and accountability in human-machine interactions, a dynamic often referred to as hybrid intelligence.

As AI systems gain autonomy, they raise complex ethical questions. For instance, if an autonomous car causes an accident, who bears responsibility, the system, the manufacturer, or the developer? Addressing these challenges demands significant human oversight to ensure AI aligns with our values and priorities. Empathy, discernment, and thoughtful governance are essential in guiding its development, use, and integration into society.

Thriving in the age of AI requires more than technical skills; it demands a philosophical shift. Building meaningful partnerships with intelligent systems involves harmonizing human creativity with machine precision and aligning AI's potential with our highest ideals.

While AI ventures into domains once considered uniquely human, like language, artistry, and problem-solving, it also highlights where humans remain irreplaceable: empathy, moral reasoning, and storytelling-sensemaking. These distinctly human traits form the foundation for a partnership that ensures our essence stays central as technology reshapes our world.

This foundational chapter highlights AI's leap beyond a technological milestone; it's a force reshaping industries, societies, and how we perceive ourselves. With this groundwork, let's explore AI's transformative reach across diverse fields and its profound implications for our future.

As we delve into the next chapter, we'll uncover how AI transcends domains, revolutionizing industries and redefining the boundaries of human potential.

Chapter 2

Embracing the AI Journey

A Tale of Aspirations, Achievements, and Realizations

"The future always seems grander in imagination than in execution. Yet, in its steady march, reality often surpasses dreams in unexpected ways."

Anonymous

Historical Overview

Artificial intelligence has long been a subject of fascination, attracting significant attention and investment throughout its development. Its history is characterized by a series of cycles, oscillating between periods of high enthusiasm and subsequent reality checks. These cycles are not merely growing pains but rather reflect the complex interaction of human psychology, market dynamics, and our collective hopes and fears about technology. By examining these cycles, we gain insight into the evolution of AI and the enduring gap between our imagined futures and the realities we manifest.

The tale of AI is one of lofty dreams, ground-breaking discoveries, and humbling setbacks. It begins with the First Wave of AI in the 1950s and 1960s, a period of boundless optimism following the Dartmouth Conference of 1956, where the term "artificial intelligence" was first introduced.

Early achievements in simple problem-solving programs and natural language processing led to bold forecasts of machines matching human intelligence within a generation. Institutions and governments generously funded this emerging field, captivated by its potential. However, as the initial enthusiasm encountered technical obstacles like the limitations of early algorithms and computing power constraints, the optimism faded. The resulting "First AI Winter" of the late 1960s to early 1970s saw a decline in funding and interest as researchers wrestled with more complex problems than initially anticipated.

With an emphasis on targeted, domain-oriented applications, the Expert Systems Era of the 1970s and 1980s rekindled optimism. Expert systems, which are intended to replicate human decision-making in domains such as chemistry and medicine, have been commercially successful and have drawn substantial corporate investment. The rise of specialized AI hardware firms demonstrated a newfound confidence in AI's commercial potential. However, this wave also met its limitations, including the high maintenance costs of rule-based systems and the emergence of more flexible software solutions, leading to another downturn. The "Second AI Winter" of the late 1980s to early 1990s once again highlighted the gap between AI's potential and its actual, practical accomplishments.

Soon after, with a more balanced approach, the machine learning revolution of the 1990s to 2010s rekindled the field. Practical applications like search engines and recommendation systems were made easier by advancements in statistical techniques and neural networks, which were bolstered by the availability of big datasets and increased processing capacity. The achievements of machine learning cleared the path for artificial intelligence's current comeback, despite going through its own usual mini-cycles of excitement and disappointment.

The current era, characterized by the Deep Learning Explosion since 2012, represents the most transformative wave yet. Advances in deep neural networks, coupled with unprecedented data availability and computational resources, have enabled AI to excel in areas once considered unattainable, including image recognition, natural language processing, autonomous systems, and more. As AI continues to permeate every aspect of society, this wave reflects not only the field's maturity but also its ongoing tension between lofty expectations and the realities of its application.

This historical journey through AI's waves of hope and disillusionment uncovers a fundamental truth: technology evolves through not just breakthroughs but also through the lessons learned from its difficulties. Understanding these cycles helps us approach AI's current state with informed optimism, recognizing its potential while acknowledging its limitations.

The Role of Media and Marketing

In the era of AI, media is not merely a conduit for information (it never is); it actively shapes perceptions of AI. Drawing on Marshall McLuhan's concept of "the medium is the message," we can see that the style, format, and delivery of AI-related stories significantly influence public perception. Dramatic images, sensational headlines, and viral social media posts magnify AI's potential and frequently present it as an all-powerful force that is poised to either transform or wipe out humanity. These mediums thrive on engagement, prioritizing bold claims and emotional resonance over nuance and accuracy. As a result, the medium of delivery reshapes the message, turning incremental technological advancements into grand narratives of triumph or catastrophe.

This dynamic is not limited to conventional media. The notion of sentient, human-like AI is deeply ingrained in our collective psyche by the entertainment industry through movies like Ex Machina and The Matrix. These cultural relics perpetuate anthropomorphic perceptions of AI, generating and perpetuating anxieties and expectations that greatly surpass its true potential. Simultaneously, tech companies, seeking investment and market dominance, craft marketing campaigns that capitalize on this hype. Buzzwords like "revolutionary," "autonomous," and "intelligent" are strategically used to portray AI as a cutting-edge phenomenon, even when referring to basic functionalities like chatbots or automated analytics.

The persistence of these narratives can be explained by McLuhan's observation that the media and marketing channels actively create AI's identity in the public consciousness rather than merely disseminating information about it. The ramifications of this phenomenon are extensive. Media narratives fuel cycles of hype and disillusionment by raising expectations to unachievable heights. Confusion, mistrust, and irrational expectations for achievement are brought about by the discrepancy between what is depicted and what is genuine.

Navigating the hype around AI requires an understanding of this interaction between medium and message. We may start to distinguish fact from fiction and appreciate AI's full potential while staying aware of its present limitations by acknowledging how media and marketing impact attitudes. In this sense, McLuhan's viewpoint offers a useful prism through which to analyze not only the statements made regarding AI but also the manner and motivation behind them.

Funding and Financial Incentives

Every wave of AI hype is powered by a potent engine of financial incentives. Startups and companies in the AI space operate in an environment where attention often translates directly into funding. Venture capitalists, corporate investors, and government agencies are drawn to technologies that promise transformative potential, pushing startups to craft narratives that align with the most sensationalized media portrayals. These dynamics create a feedback loop where funding pressures amplify the very hype that investors seek to capitalize on, perpetuating cycles of inflated expectations.

No matter how noble or visionary the initial intentions of founders may be, progress in the AI industry (and beyond) is ultimately driven by incentives, not intentions. Startups face immense pressure to secure funding, and in this high-stakes environment, intention often gives way to the need to impress investors. Buzzwords like "quantum leap," "agentic," and "transformative" dominate pitch decks, with even basic machine learning models rebranded as state-of-the-art artificial intelligence. Founders, eager to stand out in a crowded market, often lean into media-fueled fantasies of AI's limitless potential, showcasing bold ambitions that far outpace the technology's actual capabilities.

Corporate players, too, are not immune to these dynamics. Established companies seeking to maintain competitive edges often announce ambitious AI initiatives to attract investors, even when those projects are in exploratory stages. High-profile acquisitions of AI startups are frequently accompanied by press releases touting future breakthroughs, creating a perception of constant progress, even if the promised advancements are years away or fail to materialize altogether.

Government funding can also play a role. National initiatives to lead in

AI research and development often emphasize grand strategic goals, such as achieving global dominance in AI or solving societal challenges. While some of these investments may have led to genuine advancements, they also place public expectations on a pedestal, pressuring researchers to deliver dramatic results that validate the funding narrative.

The consequences of this incentive-driven ecosystem are profound. Funding, while essential for innovation, often flows disproportionately to projects that promise headline-worthy results, leaving less glamorous but equally important areas of AI research underfunded. Additionally, the pressure to deliver on inflated claims can lead to ethical oversights, rushed deployments, and public backlash when technologies fail to meet expectations.

For the AI industry to move beyond the cycles of hype and disillusionment, stakeholders need to recognize the primacy of incentives over intentions. Startups need to balance ambition with honesty, crafting narratives that inspire investors without overstepping feasibility. In turn, investors ideally discern between genuine innovation and marketing gimmicks, ensuring resources are allocated to projects with meaningful potential.

Unraveling AI's Popular Image vs. AI's Reality: Understanding AI's True Nature

Artificial intelligence, in the public consciousness, often holds a fantastical, almost mythical status. It's commonly portrayed as a powerful entity, either benevolent or dystopian, capable of outsmarting humanity. These perceptions, often fueled by the media, cultural narratives, and even ambitious marketing campaigns, are a far cry from the actual state of AI today. While AI has indeed achieved impressive feats, such as mastering complex games and generating human-like text, these achievements often lead to broad misinterpretations of AI's abilities. One widespread myth is that AI possesses sentience or the ability to think independently. High-profile examples like OpenAI's ChatGPT or Google's AlphaGo are sometimes misconstrued as evidence of human-like cognition. However, these systems are not self-aware; they are advanced pattern recognition tools that generate outputs based on the extensive datasets they've been trained on. For instance, ChatGPT does not "understand" language in the same way humans do; it predicts the most likely sequence of words based on context and statistical probabilities. Similarly, AlphaGo's success in the game of Go was not the result of intuition but of training on millions of game scenarios and using reinforcement learning algorithms to optimize strategy.

Another common misinterpretation is the overgeneralization of AI's scope. Many people assume that if AI excels in one area, such as image recognition, it can easily apply that expertise to other domains. In reality, most AI systems are currently still largely narrowly focused, designed to perform specific tasks, perhaps a series of tasks, within well-defined parameters. An AI that is trained to diagnose diseases from medical images, for example, would be ineffective in generating a marketing campaign or managing inventory for a retailer. This limitation reflects the distinction between narrow AI, where most current systems operate, and general AI, the hypothetical, human-like intelligence often depicted in fiction. These misconceptions are not trivial. They shape public expectations, influence policy decisions, and drive funding priorities, often at the expense of a more grounded understanding of AI's potential and constraints.

Overestimating AI's capabilities can lead to misplaced fears, such as the notion of machines "taking over," while underestimating its limitations fosters disillusionment when promised breakthroughs fail to appear. In order to use AI effectively, we need to debunk these misconceptions and see it for what it is: a universally empowering force that is more like electricity than a static instrument. It holds immense promise for sure, but also faces inherent danger and constraints, calling for thoughtful guidance and ethical stewardship to ensure it advances our broader societal goals.

The Tech Industry's Buzz Ecosystem

As we have seen previously, the tech industry thrives on buzz, an intricate ecosystem of buzzwords, marketing strategies, and high-profile launches designed to capture attention, attract investment, and drive adoption. In the realm of artificial intelligence, this ecosystem plays a central role in promoting and sustaining the cycle of hype, shaping not only public perceptions but also industry priorities and resource allocation. For example, relatively simple data analytics tools are frequently branded as "AI-powered," creating the impression of groundbreaking innovation where modest incremental progress is the reality. Terms like "neural networks" and "machine learning" are used liberally, often without clarifying their narrow, task-specific capabilities.

This deliberate vagueness inflates expectations, leaving customers and stakeholders with an oversized sense of AI's readiness to solve intricate and complex problems. Industry events and conferences further amplify this dynamic. High-profile showcases like the Consumer Electronics Show (CES), an annual technology trade show held in Las Vegas, or AI-dedicated expos often feature dramatic demonstrations of AI's capabilities: robots performing tasks, AI-generated artwork, or predictive models making impressive forecasts. These demonstrations, while technically impressive, are often carefully curated to highlight ideal scenarios, leaving out the limitations, dependencies, and contextual nuances that underpin the deployment of these technologies.

As a result, they contribute to a perception of AI as a fully autonomous and universally applicable force, rather than a suite of specialized collaborative tools requiring human oversight and contextual understanding. Startups, especially those in early funding stages, are particularly incentivized to ride the wave of buzz, as we have already touched upon previously. A well-crafted pitch filled with industry jargon and promises of disruptive breakthroughs can attract millions in venture capital. However, this focus on buzz often pushes companies to prioritize marketable features over robust, scalable, or ethical solutions. Overpromising and underdelivering become systemic risks, as investors and customers begin to expect results that current technology cannot provide.

The role of tech influencers and thought leaders cannot be overlooked. Inspirational remarks from high-profile figures often shape the narrative around AI's potential. Visionaries such as Elon Musk (Tesla, SpaceX), Sundar Pichai (Alphabet/Google), and Jensen Huang (Nvidia) routinely make bold predictions about AI's trajectory, driving enthusiasm but also fostering misconceptions. Amplified by social media and news outlets, these voices reinforce the cycle of hype, ensuring the buzz around AI remains loud and persistent.

While this ecosystem has its benefits, sparking innovation, attracting talent, and driving investment, it also has significant downsides. The relentless pursuit of buzz creates an environment where ethical considerations, long-term feasibility, and meaningful applications may take a backseat to marketability. Moreover, it fuels the cycles of overhype and disillusionment, as the reality of AI struggles to meet the inflated expectations generated by the buzz. Ideally, marketing would focus on educating stakeholders about what AI can and cannot do, while conferences and demonstrations would balance ambition with honest discussions of risks, challenges, and limitations.

The Role of Human Psychology

Human psychology plays a hugely significant role in perpetuating myths and exaggerations about artificial intelligence. As a species, we are wired to seek patterns, assign agency, and attribute human-like qualities to the things around us, a phenomenon known as anthropomorphism. This psychological predisposition, combined with a natural fear of the unknown, creates fertile ground for misunderstandings about AI and its capabilities. Anthropomorphism leads us to project human traits onto machines, especially those that closely mimic human behavior. When a chatbot generates coherent responses or a robot moves with human-like precision, it's easy to imagine that these systems possess understanding, emotion, or intent. Even though AI systems like ChatGPT merely predict words based on probabilities, many users describe interactions with them as though they are conversing with a sentient being. This tendency is reinforced by media and cultural portrayals, where AI is often depicted as self-aware or emotionally intelligent, further embedding anthropomorphic views into the public imagination.

Fear of the unknown also drives exaggerated perceptions of AI. As a "black box" technology, AI operates in ways that are often opaque to the average person, making it seem mysterious and potentially uncontrollable. This uncertainty fuels anxieties about AI taking over jobs, outsmarting humans, or even acting against human interests. Stories of AI systems outperforming humans in tasks like strategic games or creative writing amplify these fears, suggesting that AI is progressing toward capabilities beyond our comprehension or control. Another psychological factor is our inherent fascination with intelligence and the future. AI, as a concept, taps into deep-seated curiosity about what it means to think, create, and evolve.

This curiosity often blurs the line between speculation and reality, leading us to overestimate AI's deployable capabilities while underestimating its current limitations. The idea of AI as a superhuman force, whether benevolent or malevolent, is compelling because it speaks to our fantastical imaginations, aspirations, and fears all at once. These psychological tendencies are not new. Throughout history, humanity has reacted similarly to groundbreaking technologies, from the printing press to the steam engine. Each innovation was accompanied by both utopian dreams and dystopian fears, reflecting our complex relationship with progress. AI, however, uniquely amplifies these reactions because it mimics an attribute, intelligence, that we consider central to our identity. Understanding the psychological factors behind AI's exaggerated reputation will enable fostering a more balanced perspective.

As we strip away the layers of hype and psychology surrounding artificial intelligence, a more realistic picture begins to emerge, one that is both less fantastical and more grounded in reality. AI, while remarkable in its achievements, is not the sentient, omnipotent force of popular imagination. Yes, it is an immensely powerful but imperfect tool, akin to electricity, shaped and applied by human ingenuity, biases, and constraints. To truly understand AI's potential and limitations, we must move beyond the narratives and examine its current state: the practical applications transforming industries, the challenges it faces, and the areas where it continues to fall short. In the next chapter, we leave behind the myths and delve into the real-world AI of today, uncovering the reality behind the headlines and the opportunities that lie ahead.

Chapter 3

The Present State of AI

Unmasking Reality Beyond the Buzz

"The greatest danger in times of rapid change is not the change itself, but to act with yesterday's logic."

Peter Drucker

Discussing the current state of AI is akin to scribbling in sand. The beginning has already faded away by the time you finish your sentence. As we enter the year 2025, the realm of artificial intelligence is progressing at an astonishing rate. Today's cutting-edge models, such as ChatGPT- o1, -o3, or Sora, may soon be seen as steppingstones to even more sophisticated innovations. By the time you read this book, some of the examples in this chapter may have been superseded by more advanced tools or completely reimagined technologies. Instead of focusing on specific tools, this chapter emphasizes the categories and principles that guide their development. Consider it as a compass, assisting you in navigating the rapidly changing landscape of AI while grounding you in the more enduring technological principles that outlast any single moment of change.

Core Technological Advances: The Pillars of Modern AI – Foundation Models

Foundation models are the engines of modern artificial intelligence, driving advancements across natural language processing, computer vision, and beyond. Like a well-designed powertrain in a car, they form the core infrastructure that powers diverse applications, from healthcare diagnostics to creative content generation. Their design emphasizes multi-task learning and cross-domain generalization, making them remarkably versatile and adaptable.

Unlike earlier models built for very specific, narrow tasks, such as translation or image recognition, foundation models are trained on vast datasets spanning multiple domains. This allows them to act as adaptable frameworks that can be fine-tuned for specialized roles while remaining flexible enough to tackle emerging challenges. Foundation models offer the underlying power for a broad range of AI applications, from chatbots to driverless cars, much like an efficient engine may be modified to power anything from a sports car to a tractor.

Their scalability and adaptability have catalyzed a wave of AI-driven innovation across industries. For example, in healthcare, foundation models assist in diagnosing all kinds of diseases by analyzing diverse data types, while in the creative industries, they enable the generation of realistic art, streaming video, music, and text. Foundation models continue to be central to AI's rapid development, pushing the field into unexplored areas even as newer iterations are created at an accelerated rate.

Large Language Models (LLMs): A New Era in Natural Language Processing

Large Language Models (LLMs) are a specific subset of foundation models, representing a significant advancement in the field of natural language processing (NLP). Built on the architecture and principles of foundation models, LLMs are specifically tailored for language-related tasks. Their distinguishing feature is their ability to process and generate text that closely resembles human language, making them revolutionary tools in understanding, generating, and transforming language.

Trained on massive datasets that include books, articles, code repositories, and even social media, LLMs such as OpenAI's GPT series, Google's BERT, and Amazon's Nova models have transformed industries by excelling in tasks like language translation, content summarization, question answering, and software generation. Although LLMs' design makes use of foundation models' scalability, their emphasis on language is what distinguishes them. They can mimic human speech, handle ambiguity, understand context, and fine-tune to very specialized tasks by combining sophisticated self-attention mechanisms with vast amounts of data. If you think of foundation models as a box of all-purpose AI tools that can be used in any field, then LLMs are a specific set of those tools that can be used to solve language and communication tasks. This specialization allows LLMs to do more than just understand language; they can also generate creative content, engage in meaningful dialogue, and perform complex reasoning tasks. For instance, while early versions of GPT introduced coherent language generation, successors like ChatGPT-o1 and -o3 have evolved into conversational systems capable of producing nuanced dialogue, assisting with coding, and even serving as educational tutors.

The revolutionary aspect of LLMs lies in their ability to bridge the gap between human and machine communication. They have transformed industries by enabling seamless customer service, real-time translation, and personalized learning experiences. By building on the robust foundation of general-purpose models and focusing specifically on linguistic capabilities, LLMs have become not only groundbreaking but indispensable in modern AI applications. Their success exemplifies the broader potential of foundation models to empower specialized systems that redefine how we interact with technology.

Multi-Modal Models: Merging Sensory Inputs

Imagine trying to describe a vibrant painting to a friend using only words or composing music inspired by a silent film. These tasks require merging sensory information, a skill humans take for granted. Multi-modal AI models are bringing machines closer to this remarkable human-like ability by integrating text, images, audio, and even video into a unified architecture. These systems, such as OpenAI's DALL·E-3 with ChatGPT-4V, DeepMind's advancements, and innovative frameworks like Sora, represent a groundbreaking leap forward in how AI perceives and interacts with the world.

Multi-modal models excel in fields where single-modal systems fall short. Sora, for instance, can generate intricate visual scenes or short video clips based on textual descriptions. Imagine asking it to create a video of "a serene sunrise over a mountain lake with birds chirping" and watching the result unfold. Simultaneously, it can narrate or translate the scene in real-time, merging vision and language seamlessly. This kind of integration feels like teaching a machine not just to "see" or "read," but to "understand" like a multi-sensory human.

The real-world applications are truly astounding. Again, in healthcare, multi-modal models analyze X-rays or MRIs while cross-referencing them with a patient's medical history, providing doctors with more holistic insights. In entertainment, they enable immersive storytelling where text, visuals, and sound come together, creating entirely new formats for movies, games, and interactive media. These models orchestrate inputs from several sensory streams into a harmonic output, much like a master conductor. What makes this evolution truly exciting is the potential for new, unforeseen, unexpected, and unintended applications. Just as the invention of the smartphone redefined connectivity, multi-modal AI has the capacity to reshape fields we haven't even imagined yet. Whether assisting in real-time language translation during global collaborations or designing personalized learning experiences that adapt to individual needs, multi-modal models are not just tools; they're the foundation for a new era of exciting AI-driven creativity and functionality.

The merging of sensory inputs is not only an incremental improvement but also a beginning of a paradigm shift. By bridging the gap between modalities, multi-modal models are redefining what it means for machines to perceive and interact with the world, pushing us closer to creating AI that approaches the richness of human perception.

Scaling Effects: The Key to Emergent Capabilities

The disruptive power of modern AI isn't just about clever algorithms; it lies in its sheer scale. Like a vast library that reveals hidden historical patterns and connections when enough books are added, foundation models grow increasingly capable as they are trained on more data and utilize greater computational resources. This scaling not only refines their core abilities but also leads to emergent capabilities, skills that were never explicitly programmed but surface as a byproduct of their sheer size and extensive training. Consider the leap from GPT-2 to GPT-4.

While GPT-2 could also generate fluent text and perform basic reasoning, GPT-4 displayed the ability to write code and handle multilingual tasks with surprising proficiency. Building on this principle, ChatGPT-o1 goes even further, achieving greater specialization and accuracy in domains such as technical writing, legal reasoning, and scientific analysis. These emergent capabilities are like discovering that a chess player, after practicing enough games can also predict moves in an entirely different strategy game; it's an unexpected, yet fascinating side effect of training at scale. Scaling is not just about making AI models bigger; it makes them smarter in unexpected ways. Such emergent abilities discovered at these scales underscore the profound potential of AI to tackle problems that once seemed beyond the reach of machines, opening doors to applications we are only beginning to imagine.

Limitations: Challenges in Reasoning and Reliability

Although foundation and multi-modal models have made significant strides, they are still essentially constrained by their dependence on pattern recognition rather than actual reasoning or inference. Artificial intelligence systems are extremely good at finding patterns in the data they are trained on, but they struggle with novel, nuanced, or unclear situations that do not fall neatly into these patterns. Even though AI models like Sora and ChatGPT-o1 are remarkably adept at understanding context, they might spread false information if the data they were trained on is incomplete or flawed. Added to these limitations are ethical quandaries. AI models that are trained on extensive, unfiltered datasets can unintentionally propagate biases or stereotypes present in the data. Their capability to generate realistic text and images, although a technical marvel, raises issues about the spreading of disinformation, deepfakes, or copyright infringement. As AI systems get increasingly complicated, their internal workings are frequently hidden by their complexity, creating "black-box" models with difficult-to-decipher decision-making processes. This lack of transparency hampers efforts to identify and rectify errors, further undermining their reliability in crucial applications such as finance, healthcare, or legal decision-making. Despite the relentless progress of AI, with models like ChatGPT-o1 and Sora branching out into new areas, these limitations serve as a reminder that progress is not a straight path and is not without obstacles. Future AI generations may surpass the capabilities outlined here, but they will likely face similar challenges related to reasoning and reliability.

Transformer Dominance: The Rise of the Attention Mechanism

The advent of Transformers brought about a paradigm shift in sequence processing through self-attention, allowing models to concentrate on different parts of an input sequence all at once. The groundbreaking 2017 paper "Attention Is All You Need" by Vaswani et al. not only introduced the self-attention mechanism but also laid the groundwork for the comprehensive Transformer architecture, which quickly redefined how large-scale AI models are built. Unlike Recurrent Neural Networks (RNNs), which process data step by step and seriously struggle with long-range dependencies, Transformers can instantly focus on any part of the sequence. This parallelism significantly speeds up training and enables Transformers to handle much more complex and lengthy sequences without losing context. Imagine listening to a lively group conversation: RNNs would process each speaker in strict order, one at a time, while Transformers can "listen" to everyone at the same time, picking up the most relevant details as needed. This ability to grasp the bigger picture while focusing on specifics makes Transformers incredibly efficient and versatile. The impact of this innovation has been nothing short of revolutionary. By eliminating the bottleneck of sequential processing, Transformers have scaled seamlessly to handle enormous datasets and longer input lengths, a critical advantage in fields where understanding context is essential, such as natural language processing, genomic research, and protein folding.

The remarkable adaptability and scalability of Transformer design are responsible for the success of models such as GPT, BERT, and others. From enabling real-time language translation to uncovering the mysteries of biological structures, Transformers have become the gold standard for tackling increasingly complex real-world challenges. Their versatility has pushed the boundaries of what AI can achieve, cementing their role as a cornerstone of modern artificial intelligence.

Superior Scalability and Parallelization

One of the defining strengths of Transformers is their ability to leverage modern hardware, such as GPUs and TPUs, for parallel processing. Unlike older models that rely on sequential data processing, Transformers perform matrix multiplication across large grids of numbers simultaneously, determining precisely which parts of the input deserve the most "attention." This efficiency allows Transformers to train faster and handle significantly larger datasets and models, unlocking unprecedented levels of superior performance. Even in tasks, traditionally dominated by Convolutional Neural Networks (CNNs), like image recognition, Transformers have proven their adaptability. Instead of analyzing every pixel or section individually, Transformers process chunks of an image in parallel, akin to a team of people examining a large mural simultaneously. This approach not only accelerates the process but also captures intricate details that a single sequential method might overlook. These advancements highlight why Transformers have become the architecture of choice for tackling diverse and challenging real-world problems, demonstrating a level of flexibility and scalability unmatched by any of their predecessors.

Continual Advancement

The Transformers discussed are far from being static entities. They are consistently evolving through enhancements such as Sparse Attention, Longformer, and Perceiver. These developments aim to further optimize how attention is calculated over extensive sequences. There is also ongoing research aimed at making Transformers more efficient by reducing their high memory and compute requirements. Looking to the future, Transformers seem likely to continue dominating a broadening spectrum of use cases, including text generation, computer vision, robotics, and more, as research continues to make them lighter, faster, and more capable of reasoning.

Few-Shot and Zero-Shot Learning: Unlocking New Possibilities

Few-shot and zero-shot learning further exemplify the transformative capabilities of foundational models. Traditional machine learning systems require vast amounts of labeled data to perform well, but foundational models, powered by their scale and adaptability, can manage tasks with minimal or no direct training examples. Few-shot learning enables AI to generalize from just a handful of examples, for instance, identifying a rare bird species after seeing only a few images. Zero-shot learning takes this even further, allowing AI to perform entirely new tasks by drawing on patterns and knowledge from similar contexts without explicit training.

These advanced learning techniques significantly reduce the resources required for gathering and labeling data, making AI more accessible for specialized or resource-constrained applications. In healthcare, for example, zero-shot learning can assist in diagnosing rare medical conditions without the need for vast diagnostic datasets. In manufacturing, it can identify unique defects in specialized processes where labeled data may be scarce or expensive to obtain. Enabling rapid adaptation to new challenges, few-shot and zero-shot learning showcase the unparalleled power of foundational models to generalize across domains. This capability not only lowers barriers to AI adoption but also highlights the broader potential of these models to address previously inaccessible problems efficiently as well as cost-effectively.

Advances in Reinforcement Learning

A fascinating branch of artificial intelligence called reinforcement learning (RL) teaches a computer program, or "agent," to make decisions by interacting with its surroundings and getting rewards or penalties for its actions. In order to maximize rewards, the agent gradually improves its conduct through trial, feedback, and adjustment. Similar to how a beginner chess player learns strategy by playing several games, losing, and changing their tactics after each match, this learning process mimics real-life situations. Two well-known achievements illustrate RL's transformational potential:

* AlphaFold (DeepMind): Revolutionizing protein structure prediction, AlphaFold employs RL-like techniques to explore vast possibilities and accurately predict the 3D structures of proteins. Imagine trying to solve a 3D jigsaw puzzle with thousands of pieces but no clear image to guide you; this is akin to the challenge AlphaFold overcomes. Its discoveries are broadening our understanding of life processes and paving the way for advancements in drug discovery.

* OpenAI Five: In the realm of competitive gaming, OpenAI Five mastered the complex, real-time strategy game Dota 2. By playing millions of matches against itself, the agent learned intricate tactics, teamwork, and adaptability, surpassing human capabilities in this multifaceted environment. This achievement is comparable to a beginner mastering a new game's rules and then coming up with creative ways to beat more seasoned players.

Reinforcement learning is transforming what machines can accomplish in dynamic and unpredictable settings, ranging from optimizing warehouse logistics to refining financial trading strategies. Its ability to adapt and learn without explicit supervision makes it a powerful tool in scenarios where predefined rules are impractical. As RL continues to evolve, its potential applications are expanding, promising revolutionary solutions to some of the world’s most complex challenges.

Robotics and Humanoid Robots

Robots are no longer confined to the predictable environments of manufacturing floors. They have expanded into diverse and dynamic fields such as healthcare, retail, and even personal companionship, heralding a new era of human-robot interaction. Humanoid robots, in particular, have captured public imagination with their ability to mimic human gestures, expressions, and interactions. Designed to seamlessly integrate into homes and workplaces, these robots aim to serve as personal assistants, caregivers, and yes, even made-to-order entertainers. For instance, robots equipped with advanced AI and sensors can recognize faces, interpret emotions, and follow complex instructions, allowing them to engage in meaningful conversations and provide personalized support. Imagine a robot assisting an elderly person by preparing meals, reminding them to take medication, or even providing social interaction in moments of loneliness. Similarly in retail, humanoid robots guide customers through stores, answering questions and providing a futuristic shopping experience.

While these developments understandably spark substantial enthusiasm, they also underscore significant technical and ethical challenges. For instance, ensuring safety and reliability in real-world scenarios is paramount, particularly when robots interact with vulnerable populations. Furthermore, the collection and processing of sensitive data by robots for task performance raises significant privacy concerns. Emotional attachment, too, raises questions about the psychological impact of forming bonds with machines. Can these relationships be meaningful, or do they risk fostering dependency and isolation?

Autonomous Vehicles and Drones

Autonomous vehicles and drones are at the forefront of technological innovation, with the potential to redefine transportation, logistics, and public services. Companies such as Tesla, Uber, Waymo, and others are testing self-driving cars and trucks on public roads with increasing confidence. These vehicles use a combination of sensors, cameras, LiDAR (a remote sensing method that uses light in the form of a pulsed laser to measure distances), and AI algorithms to detect and respond to their surroundings with extreme accuracy in real-time. The promise of these systems extends beyond convenience, offering improved safety by reducing human error, currently the leading cause of traffic accidents. Likewise, in commercial settings, autonomous trucks are being trialed for long-haul freight, where they could operate continuously without ever taking rest breaks, dramatically reducing delivery times and operational costs. In select locations, self-driving taxis are being used to demonstrate the possibility of a future in which driverless, on-demand transportation is commonplace, enhancing urban mobility and easing traffic.

Drones, meanwhile, are proving their value in both commercial and public service roles. In logistics, companies like Amazon are piloting drone delivery systems capable of transporting packages directly to customers' doorsteps, particularly in rural or hard-to-reach areas. In agriculture, drones monitor crops, optimize irrigation, and detect signs of pests or disease, enabling precision farming that saves precious resources and boosts yields. Emergency services are also leveraging drones for disaster response, providing aerial surveillance to locate survivors, assess damage, and deliver critical supplies in areas inaccessible to traditional vehicles. What makes these technologies particularly transformative is their potential to operate sustainably.

Autonomous vehicles powered by green energy, such as electric or hydrogen fuel cells, could significantly reduce emissions compared to traditional combustion engines. Also, drones, with their small size and efficiency, offer a low-impact solution for last-mile delivery, further reducing the carbon footprint of logistics operations. As these technologies advance, they present not only opportunities but, as expected, critical questions about their integration into society. Issues such as cybersecurity, air traffic management for drones, and the displacement of human jobs require some thoughtful solutions. Nevertheless, autonomous vehicles and drones are undeniably driving a future where transportation and delivery systems become safer, more efficient, and increasingly eco-friendly. Their progress signals an unstoppable shift that could reshape industries and everyday life alike.

Remaining Hurdles and Balancing Hype with Reality

Indeed, the vision of a future filled with self-driving cars, helpful humanoid companions, and drones delivering packages to our doorsteps is captivating. However, significant challenges remain on the road to widespread adoption of autonomous systems. Translating complex, dynamic real-world environments into data that machines can accurately process is a persistent hurdle. Errors in sensing or decision-making can lead to accidents or malfunctions, emphasizing the critical need for robust testing, fail-safe mechanisms, and continuous algorithm refinement.

For autonomous vehicles, "edge cases," unpredictable scenarios like erratic pedestrians, sudden road closures, or extreme weather conditions, pose ongoing challenges. These rare but critical events highlight the limitations of current systems and the necessity of further advancements to ensure safety and reliability. Similarly, drones face unique obstacles, including stringent regulatory constraints, safety concerns over crowded areas, and limited battery life that restricts their operational range. Addressing these issues will require breakthroughs in battery technology, air traffic management, and adherence to evolving and ever-increasing regulatory frameworks.

Beyond the technical barriers, there are significant ethical, legal, and societal questions to consider. Who bears responsibility in the event of an accident involving autonomous systems? How do we safeguard privacy and prevent misuse, such as unauthorized drone surveillance? Also, we need to think carefully about the psychological and social effects of living with machines that are becoming more and more autonomous, especially in fields where people may feel displaced or overshadowed.

Looking ahead, the story of AI is not just about its transformative applications but also about the foundation that makes these advancements possible. In the next chapter, we delve into the hardware and software backbone of AI, exploring the cutting-edge technologies that power its capabilities. From GPUs and TPUs to cloud computing and AutoML, we'll uncover how this infrastructure supports the scalability, efficiency, and innovation driving AI's rapid evolution. Understanding this foundational layer is key to appreciating the breakthroughs shaping industries and society.

Chapter 4

The Machinery of Minds: Hardware and Software Driving AI

The Infrastructure of Intelligence

"The tools we use shape the way we build, and the foundation determines the height of our aspirations."

Anonymous

Introduction to the Technical Foundations of AI

Artificial Intelligence is transforming industries and reshaping the world, but not everyone needs to dive deep into the mechanics to harness its power. This chapter is designed for those who are more technically inclined, those curious about the nuts and bolts of the AI ecosystem. Here, we break down the essential hardware, software, and optimization techniques that drive modern AI systems.

However, if you're more interested in the practical applications of AI rather than its inner workings, feel free to move on to the next chapter. You don't need to understand every technical detail to thrive in an AI-driven world. What matters most is knowing how to leverage AI effectively, whether as a tool for personal growth, business success, or creative exploration. But for those eager to peek under the hood, let's dive in.

Hardware Foundations: Fueling the AI Revolution

The transformative power of AI is rooted in its hardware, the machinery that propels computation on a scale that was unthinkable just a few decades ago. The evolution of these systems has been instrumental in propelling AI's rapid advancement. Initially, CPUs (Central Processing Units) were the pillars of computing, efficiently handling general-purpose tasks. However, as AI models increased in complexity, the demand for quicker and more specialized processing emerged. This led to the introduction of GPUs (Graphics Processing Units), designed for parallel processing, which revolutionized the training of deep learning models. For instance, NVIDIA's GPUs facilitated the training of models like AlexNet, heralding the advent of modern deep learning. Today, specialized hardware like TPUs (Tensor Processing Units) and NPUs (Neural Processing Units) are pushing the boundaries even further. Models like BERT and GPT are powered by Google-developed TPUs, which offer unmatched speed and energy efficiency. Meanwhile, real-time picture recognition and other AI capabilities are now within our reach thanks to NPUs, which are integrated into smartphones. With these developments, general-purpose units are giving way to task-specific ones that are better suited for AI workloads.

Memory and Storage: The Silent Heroes

While processors often take center stage in AI performance, memory and storage play an equally critical role in handling vast amounts of data efficiently. AI workloads demand rapid data retrieval, seamless processing, and high throughput, all of which depend on specialized memory and storage solutions.

* High-Bandwidth Memory (HBM): Enables ultra-fast data transfer between processors, crucial for deep learning models that process massive datasets in real-time.

* Solid-State Drives (SSDs): Provide the high-speed storage required for managing and retrieving large AI training datasets efficiently, significantly reducing data bottlenecks.

* RAM (Random Access Memory): Facilitates real-time data processing by temporarily storing active computations, ensuring smooth AI model train-

ing and inference.

Emerging technologies like in-memory computing aim to minimize the latency associated with data movement between storage and processors. By processing data directly in memory, these advancements reduce power consumption and accelerate AI workloads, paving the way for more efficient AI-driven systems.

Software Ecosystems: The Brains Behind the Hardware

Just like a driverless car needs software to function, AI hardware relies on powerful software frameworks to bring it to life. Over the past decade, advancements in AI software have democratized access to AI development, making innovation more widespread and accessible.

Development Frameworks

Deep learning frameworks like TensorFlow and PyTorch provide essential tools for building, training, and deploying AI models. These frameworks simplify complex tasks, allowing developers to focus on inovations rather than technical obstacles. PyTorch, for example, features dynamic computation graphs that enable more flexible experimentation with new AI architectures. Platforms like Hugging Face have further revolutionized AI development by offering pre-trained models and user-friendly tools, making AI more accessible across industries, from startups to multinational corporations.

AutoML: Democratizing AI

Automated machine learning (AutoML) platforms help businesses and organizations develop AI solutions without requiring deep technical expertise. Platforms such as Google Cloud AutoML, Azure AutoML, and Amazon SageMaker Autopilot automate key steps like data preprocessing, model selection, and optimization. This allows companies to leverage AI for real-world applications with minimal coding. For example, a retailer can use AutoML to build demand forecasting models that account for trends, seasonality, and local events, while their data teams focus on strategy and insights. However, successful AI implementation still requires an understanding of data quality, business needs, and responsible AI governance.

Optimizing for efficiency

As AI models grow more complex, optimization techniques are essential for making them practical and efficient. These methods focus on improving model structure, reducing memory usage, and increasing computational efficiency.

* Model Pruning: Removes less important neural connections while preserving accuracy, reducing model size and speeding up inference.

* Quantization: Converts high-precision weights (e.g., 32-bit floating-point) to lower-precision formats (e.g., 8-bit integers), cutting memory usage and boosting performance.

* Knowledge Distillation: Transfers learning from large "teacher" models to smaller "student" models, making them more efficient while retaining much of their original capability.

These techniques are especially valuable in edge computing, where AI must run on devices with limited processing power. For example, a smart home device might use a pruned and quantized speech recognition model to process voice commands locally, achieving fast responses while protecting user privacy.

Other advanced methods include neural architecture search, which automates the design of efficient models, and hardware-aware optimization, which tailors models for specific platforms like mobile processors or specialized AI accelerators. These strategies help ensure AI can run effectively across diverse environments, from cloud servers to compact edge devices.

Monitoring and Deployment

Once an AI model is trained and optimized, the next step is deployment and ongoing monitoring to ensure reliable real-world performance. Models are typically deployed via REST APIs (Representational State Transfer Application Programming Interface) for on-demand inference or streaming services for continuous data processing. Deployment strategies range from serverless functions, which offer cost-effective scalability, to containerization tools like Docker, which simplify environment management. For large-scale applications, Kubernetes provides orchestration, ensuring models run reliably across on-premises or cloud-based infrastructure.

A global e-commerce platform, for instance, might implement a recommendation engine using a hybrid approach: real-time inference to personalize user experiences instantly and batch processing for daily model updates, dynamically scaling resources during peak traffic periods.

Managing the full model lifecycle requires an MLOps (Machine Learning Operations) toolchain, including tools like MLflow, Kubeflow, and Weights & Biases. These provide essential features such as model versioning, experiment tracking, automated deployment pipelines, and performance monitoring.

Key metrics tracked by monitoring systems include:

* Resource utilization (CPU, GPU, memory usage)

* Data drift (shifts in input data distributions)

* Inference latency (response time for model predictions)

* Prediction accuracy (ongoing evaluation of model correctness)

Advanced monitoring solutions can detect shifts in feature importance or data relationships, helping teams anticipate performance degradation. This proactive maintenance ensures models remain accurate and reliable, enabling teams to update, retrain, or optimize models as needed to adapt to changing environments.

The Emergence of Cloud Computing and Amazon's Role

The story of cloud computing begins with Amazon. In the early 2000s, Am azon.com was grappling with significant infrastructure challenges. During periods of high shopping demand, its online store needed a large amount of server capacity, while at other times, the hardware was frequently idle. Amazon created an internal method to solve this problem, enabling engineers to swiftly build up and disassemble virtual servers as needed. Understanding the business potential of this strategy, Amazon made Amazon Web Services (AWS) available to the public in 2006. This move enabled companies of all sizes to rent Amazon's computing power, rather than purchasing and maintaining physical servers, thereby revolutionizing the IT landscape. Over time, AWS has evolved into a vast ecosystem that includes storage (S3), computing services (EC2), databases (RDS), and increasingly,

AI-specific services such as Amazon SageMaker.

Cloud as Infrastructure, Platform, and Service

When we refer to "the cloud," we're actually discussing different levels of computing resources, each catering to distinct user needs:

* Infrastructure as a Service (IaaS): This is the most fundamental level of the cloud, where providers like AWS, Microsoft Azure, or Google Cloud offer on-demand virtual machines, storage, and networking. You manage the operating systems, applications, and data; the cloud provider takes care of the underlying hardware.

* Platform as a Service (PaaS): At this level, providers offer a more comprehensive environment for development and deployment. You don't have to worry about configuring servers or operating systems. Instead, you focus on writing code, and the platform handles runtime, libraries, and other dependencies.

* Software as a Service (SaaS): This is the highest-level offering, where end-users access the software via a web browser or an app (e.g., Google Workspace or Salesforce). The provider manages both the infrastructure and the platform, while customers simply use the software. In the AI domain, Amazon SageMaker (PaaS) is a prime example: you don't need to set up your own infrastructure from scratch; instead, SageMaker provides managed notebooks, training clusters, and deployment endpoints for sophisticated machine learning tasks.

Constructing, Implementing, and Sustaining Sophisticated Architectures

One of the primary benefits of AWS (and similar platforms) is the ability to construct complex solutions by integrating multiple services. For example, an AI workflow could store extensive training data in Amazon S3, utilize GPU-powered EC2 instances for rigorous model training, implement the resulting model on SageMaker for real-time inference, and employ AWS Lambda functions to manage event-driven processes.

Security measures such as CloudWatch, CloudTrail, and Identity and Access Management (IAM) guarantee that these intricate configurations are both reliable and secure. These architectures are modular and scalable, allowing you to augment computing power during high-demand periods or scale back when demand declines. Moreover, these services run on a "pay-as-you-go" model, which is particularly attractive to smaller organizations that are unable or unwilling to make upfront investments in costly hardware.

Empowering Smaller Organizations

One of the significant outcomes of the cloud revolution is the democratization of advanced technology. Prior to the creation of AWS, only large corporations with substantial budgets could afford supercomputers or extensive server farms. Today, a startup can rent the same high-performance infrastructure for as long as it requires. This accessibility fosters experimentation and innovation, allowing smaller entities to develop sophisticated AI applications without the cost and complexity of building their own data centers. By leveraging the cloud as IaaS, PaaS, or SaaS, organizations of all sizes can access powerful computing resources, cutting-edge AI frameworks, and managed services. Whether the goal is to create a new machine learning tool or rapidly deploy a customer-facing AI solution, cloud platforms today provide the necessary components, allowing companies to focus on innovation rather than infrastructure.

Edge Computing: Local AI for Speed and Security

The practice of running AI models directly on a device, such as a smartphone, a smart camera, or an autonomous drone, is known as edge computing. Instead of sending data to a remote data center for processing, devices perform computations locally. Several factors contribute to the shift towards edge computing, including:

* Reduced Latency. Devices no longer need to engage with a remote server, resulting in quicker response times. For instance, a smart doorbell that recognizes a familiar face can unlock the door almost instantly without waiting for a round-trip to the cloud.

* Enhanced Privacy and Security. Storing data on the device means personal or sensitive information does not leave your local network. This minimizes the risk of interception or misuse of data and can help organizations comply with stricter data-privacy regulations.

* Offline Capabilities. Edge devices can retain some functionality even when not connected to the internet. A smartphone equipped with an

on-device translation model can still translate speech on-the-go, without relying on a network connection. For instance, consider how newer smartphones allow you to use voice commands (like "Open my calendar") without transmitting your speech to a remote server. The AI model, often referred to as a speech recognition engine, resides on the phone itself.

Edge computing technology allows your device to process your voice input immediately, saving time, safeguarding your data, and ensuring the feature works even if you're offline.

Energy Challenges: The Environmental Impact of Large-Scale AI

Training sophisticated AI models is incredibly energy-intensive, often requiring enormous computing power operating continuously for days or weeks. This can lead to significant carbon emissions, particularly when data centers rely on non-renewable energy sources. Consider this: Training a state-of-the-art language model can consume as much energy as powering a small town for the same duration, raising important ethical questions about balancing technological innovation with environmental responsibility. In response, leading tech companies and research institutions are adopting more sustainable practices. Some are relocating data centers to regions rich in renewable energy, such as hydroelectric or wind power. Others are experimenting with methods like model distillation (reducing model size without sacrificing too much accuracy) or sparsity techniques (removing unnecessary connections in neural networks) to reduce computational overhead. Additionally, data centers themselves are being optimized for energy efficiency, using advanced cooling systems and AI-driven resource allocation to minimize waste. Ultimately, while AI's demand for electricity remains substantial, ongoing efforts aim to ensure these powerful technologies evolve in tandem with responsible, eco-friendly solutions.

Bringing It All Together

A practical scenario might involve a team using PyTorch to build a custom object-detection model, relying on AutoML services to quickly validate architecture choices. They distill the final model for edge devices, then deploy it via containerized microservices on a cloud platform. Ongoing monitoring with MLflow ensures the team can manage updates efficiently. This end-to-end pipeline demonstrates how each tool contributes to a seamless AI development lifecycle, making advanced machine learning workflows more accessible and more powerful than ever before.

As we've explored in this chapter, the hardware and software foundations of AI form the essential backbone that powers its remarkable capabilities. However, the true impact of AI lies not only in its underlying infrastructure, but also in the application of these technologies to solve real-world problems.

In the next chapter, we'll step out of the technical realm and into the industries where AI is transforming the way we work, live, and interact. From revolutionizing healthcare and finance to reshaping education, manufacturing, and beyond, we'll uncover the innovative ways businesses and organizations are leveraging AI to drive efficiency, unlock new opportunities, and meet the challenges of a rapidly changing and increasingly demanding world.

Chapter 5

The Transformative Impact of AI Across Industries

Practical Applications and Innovations

"The art of progress is to preserve order amid change and to preserve change amid order."

Alfred North Whitehead

Enterprise Adoption

Artificial intelligence has surpassed its beginnings in research labs to become a transformative force that is reshaping industries worldwide. No longer confined to theoretical potential, AI is now the engine driving innovation in businesses of every size and sector. From predicting customer needs with pinpoint accuracy to optimizing entire supply chains, organizations are unlocking unprecedented possibilities.

In healthcare, AI is diagnosing diseases with remarkable precision, while in retail, it personalizes shopping experiences that feel tailor-made. Financial institutions rely on AI to detect fraud in real time, and manufacturers are using intelligent robots to streamline production lines. Beyond automation, AI is enabling a new era of strategic thinking, empowering companies to make faster, smarter decisions backed by powerful, data-driven analytics.

As enterprises embrace this technology, the impact is evident: improved efficiency, significant cost savings, and enhanced customer experiences. Yet, the path to adoption is not without its challenges. In this chapter, we delve into how industries are integrating AI into their operations, the transformative results they are achieving, and the lessons they're learning along the way. This is the story of AI's leap from innovation to industry revolution, driving change not just in technology, but in how businesses compete, innovate, and thrive.

Business Process Automation: Reinventing Efficiency

In the age of cloud computing and AI, businesses are no longer limited to fragmented solutions for operational workflows. Cloud-based AI integration platforms are revolutionizing how companies automate and streamline their processes. These platforms enable seamless collaboration between AI-powered tools, legacy systems, and cloud applications, automating entire operational workflows end-to-end. Large consultancy firms like PwC, Deloitte, and Accenture play a pivotal role in this transformation, guiding businesses in leveraging these platforms to modernize their operations and unlock their full, previously untapped potential.

One of the most impactful areas is Business Process Automation (BPA), where repetitive administrative tasks like data entry, invoice processing, and report generation are handed over to AI-driven tools. Robotic Process Automation (RPA) software, for instance, employs "bots" that mimic human interactions with existing systems and applications. Imagine a bot logging into a database, extracting relevant data, populating a spreadsheet, and generating an invoice; tasks that previously took hours or even days for entire teams to complete can now be performed in minutes.

Cloud-based platforms take this automation a step further by integrating RPA with AI capabilities like natural language processing (NLP) and predictive analytics. For example, an AI system integrated with a company's email and document storage can automatically process customer inquiries, draft responses, and even flag potential contract risks for human review. In industries such as insurance, AI-powered workflows can assess claims, verify documentation, and initiate payouts, creating faster resolutions and enhancing customer satisfaction.

This change not only lowers operating expenses but also radically alters how companies distribute their resources. By automating routine tasks, employees are freed to focus on strategic, high-value work, driving innovation and creativity. For instance, accountants in the financial industry can focus on strategic analysis instead of laborious reconciliation, and recruiters in human resources can focus on fostering relationships rather than reviewing resumes.

As businesses increasingly adopt AI-powered automation, the role of consultancies becomes critical in ensuring smooth integration, addressing challenges like data compatibility, security, and workforce adaptation. These consultancies help businesses bridge the gap between strategy and technology by facilitating the shift from conventional workflows to AI-driven ecosystems that are not just leaner and faster but also more adaptable to new developments.

Customer Service Transformation: Redefining the Customer Experience

Conversational AI agents and virtual assistants are redefining the way businesses connect with their customers, offering a level of responsiveness and personalization previously unimaginable. These AI-driven tools handle everything from answering product inquiries to troubleshooting technical issues, providing 24/7 support that eliminates long wait times and enhances the customer experience. AI chatbots, for instance, help customers in the financial industry with tasks like changing passwords, checking account balances, and even providing personalized financial advice based on their spending patterns and objectives.

These AI solutions make it possible for human support teams to concentrate on more complex and sensitive issues by effectively handling large numbers of regular queries. This change promotes a more human-centered approach to problem-solving where it counts most while reducing operational bottlenecks. Through intelligent automation, businesses and customers alike benefit from streamlined processes, lower costs, and increased customer satisfaction.

Decision Support Systems: Empowering Smarter Leadership

AI-powered decision support systems are transforming strategic planning and crisis management, equipping executives with the tools to navigate complex challenges and seize emerging opportunities. These systems look at huge amounts of data, like market trends and supply chain metrics, and come up with useful insights that help leaders predict risks, make the best use of resources, and feel more confident in the decisions they make based on data.

For example, an AI platform might recommend real-time adjustments to manufacturing schedules when material shortages arise or offer scenario-based forecasts to help retailers understand how potential policy changes could impact revenue. These systems excel at processing and interpreting information at a scale and speed unattainable by human teams alone, presenting leaders with clear, actionable intelligence.

With AI handling the heavy lifting in data analytics, decision-makers are freed to focus on applying their domain expertise and strategic vision to shape the organization's trajectory. The result is a powerful synergy between advanced technology and human insight, enabling companies to respond to challenges more proactively and steer confidently into the future.

Predictive Analytics and Maintenance: Proactively Preventing Downtime

One of the most impactful applications of AI lies in preventing costly equipment failures through predictive maintenance. In industrial settings like factories or energy plants, unplanned downtime caused by machinery breakdowns can lead to significant financial losses. Predictive maintenance systems leverage data from sensors and historical performance records to identify patterns that signal potential malfunctions.

For instance, when equipment begins to exhibit early signs of wear and tear, AI-powered systems can alert service technicians before the problem escalates into a critical failure. This proactive approach enables timely repairs, minimizing downtime and avoiding the high costs associated with emergency breakdowns. Maintaining machinery in optimal condition, businesses not only enhance operational efficiency but also extend the lifespan of their assets, translating to substantial savings in repair and replacement costs.

Predictive analytics not only improve maintenance; it transforms it into a strategic advantage, empowering companies to stay ahead of problems and focus on maximizing productivity and profitability.

Governmental Adoption of AI: Transforming Public Services and Governance

Governments around the world are embracing AI to enhance efficiency, improve decision-making, and deliver better public services. Much like in the private sector, AI technologies are revolutionizing how governmental agencies manage resources and address societal needs. By analyzing vast datasets from sectors such as healthcare, education, public safety, and infrastructure, AI systems can pinpoint funding priorities, predict emerging needs, and help policymakers make more informed decisions.

Across many jurisdictions, AI-powered platforms have simplified bureaucratic processes, accelerating citizen requests and cutting through administrative backlogs. Virtual assistants are increasingly being deployed to handle routine inquiries about taxes, social services, and other public matters, making government interactions more accessible, transparent, and efficient. For example, an AI assistant might guide a citizen through filing a tax return or accessing housing benefits, significantly reducing the time and frustration typically associated with such tasks.

AI's potential extends to national security and public safety. Law enforcement agencies are using AI-driven analytics to identify patterns in criminal activity, optimize policing strategies, and allocate resources more effectively. In emergency management, AI systems analyze weather data, infrastructure vulnerabilities, and real-time conditions to predict natural disasters, coordinate emergency responses, and protect lives. These capabilities have made AI an indispensable tool in safeguarding communities and responding to crises.

However, the adoption of AI in government raises critical ethical and practical challenges. Public-sector AI often relies on citizens' data, making issues like data privacy, fairness, and accountability paramount. Missteps or errors in these areas could erode public trust and undermine the effectiveness of AI-driven initiatives. Ensuring that AI systems are transparent, equitable, and subject to robust oversight is essential to maintaining public confidence. Moreover, governments must address inherent biases in AI algorithms, especially when these systems impact marginalized communities or sensitive areas such as law enforcement.

Striking a balance between innovation and responsible governance is key to unlocking AI's full potential in the public sector. By embracing these technologies thoughtfully and ethically, governments can not only enhance public services and resource management but also set a global standard for the responsible use of AI in protecting and serving their citizens.

Innovations in Specific Sectors

Healthcare: AI as a partner in medicine

In the words often attributed to Voltaire, "The art of medicine consists in amusing the patient while nature cures the disease." Despite the centuries that have passed since Voltaire's time, the sentiment remains relevant in today's AI-driven healthcare environment. Used appropriately, artificial intelligence has the potential to free doctors from routine tasks, allowing them to focus more on connecting with their patients, providing reassurance, empathy, and personalized care.

AI is transforming healthcare by acting as a powerful ally to medical professionals, enhancing their capabilities rather than replacing them. Imagine an AI-powered radiology tool scanning a chest X-ray, identifying subtle signs of disease that might otherwise go unnoticed. Within moments, it flags potential issues, equipping doctors with actionable insights for quicker and more accurate diagnoses. These tools, powered by machine learning, excel at analyzing medical images like X-rays and MRIs, detecting early indications of conditions such as cancer or heart disease with remarkable speed and precision.

Beyond diagnostics, AI is revolutionizing drug discovery. Startups like Atomwise use machine learning to predict molecular interactions at the atomic level, enabling researchers to identify promising drug candidates faster and at significantly lower costs. This capability accelerates the development of life-saving treatments, bringing hope to patients much sooner than traditional methods allow.

AI relieves doctors of the burden of routine and technical tasks, allowing them to concentrate on their most human role: being present for their patients. It paves the way for personalized medicine, tailoring therapies to individual needs while improving outcomes. These advancements are not only increasing the effectiveness of healthcare but also fundamentally changing the way we deliver care and returning doctors to their primary role as companions, healers, and recovery mentors.

Finance: Redefining the Rules of Money Management

Financial institutions have been at the forefront of AI adoption, leveraging its power to enhance security, optimize investments, and make smarter lending decisions. Think of a fraud detection system that sifts through millions of transactions every second, spotting unusual patterns, like a credit card being used in two countries simultaneously, that might otherwise slip past human analysts. These AI algorithms not only identify potential fraud faster but also help prevent losses before they escalate.

In the fast-paced world of trading, algorithmic systems powered by AI are transforming how investments are made. By analyzing historical data and market trends, these systems can execute split-second trades that capitalize on fleeting opportunities, such as price changes triggered by breaking news. For example, during a surprise policy announcement, an AI model could instantly adjust a portfolio to minimize risks or exploit sudden market shifts; moves far too fast for a human to achieve.

AI is also reshaping credit scoring. Traditional methods often rely on narrow criteria like credit history, which can exclude people with non-traditional financial backgrounds. AI-driven tools, however, evaluate a broader range of data points, such as income trends, utility payments, or even educational background, to assess creditworthiness more comprehensively. This allows financial institutions to extend credit to underserved populations while responsibly managing risk, making borrowing fairer and more inclusive.

From catching fraud in real-time to democratizing access to credit, AI is redefining the way money moves and is managed, ideally, making finance smarter, safer, and more accessible for businesses and individuals alike.

Manufacturing: Precision, Efficiency, and Smarter Supply Chains

AI is revolutionizing manufacturing by bringing precision, consistency, and intelligence to every stage of the production process. Imagine a car assembly line where AI-powered cameras and sensors inspect each component in real time, spotting minute defects that might escape the human eye, a tiny crack in a windshield or a misaligned part. By identifying issues early, these systems ensure higher quality standards, reducing costly recalls and protecting brand reputation.

Smart robotics elevate automation by effortlessly managing repetitive tasks with unwavering speed and precision. For example, in electronics manufacturing, robots equipped with AI can assemble delicate components like circuit boards with a level of accuracy that ensures consistent performance. Unlike human workers, these robots don't tire under heavy workloads, maintaining efficiency around the clock and increasing overall productivity.

Beyond the factory floor, AI is transforming supply chain management into a proactive, intelligent system. Using real-time data analytics, AI can predict demand surges, anticipate delays, and dynamically allocate resources. Picture a manufacturer using AI to forecast a spike in demand for electric vehicles and preemptively ramping up battery production, avoiding supply shortages while reducing waste. In logistics, AI can optimize delivery routes and schedules, ensuring materials arrive on time and minimizing downtime.

From defect detection to dynamic supply chain adjustments, AI is not just supporting manufacturing; it's enabling a new era of smarter, leaner, and more responsive industrial processes. These innovations are driving the industry toward greater efficiency, reliability, and adaptability in an increasingly competitive global market.

Education: Redefining Learning for the AI Era

AI is also redefining education by creating dynamic, individualized learning environments in both online and traditional classroom settings. Consider AI-powered platforms like Khan Academy's AI extensions or Coursera's adaptive learning tools, which tailor lessons to individual students. A math student struggling with algebraic equations, for instance, might receive additional examples, step-by-step explanations, and targeted practice problems, while an advanced learner is seamlessly guided to more challenging topics. This real-time adaptation allows students to learn at their own pace, bridging gaps in understanding and accelerating proficiency.

For educators, AI offers transformative tools that enhance teaching.

Real-time analytics identify knowledge gaps across an entire class, enabling instructors to intervene early and provide focused support where it's most needed. Platforms like EdTech AI offer dashboards that show which concepts students struggle with most, allowing teachers to adjust lesson plans accordingly. Routine tasks like grading multiple-choice quizzes or sorting homework submissions are automated, freeing teachers to focus on interactive, high-value activities such as facilitating discussions, mentoring, or fostering creativity.

Entirely new ways of education are emerging, challenging legacy institutions to rethink the very concept of learning. Platforms like Duolingo, which gamify language learning with AI-driven feedback and rewards, or BYJU's, which uses AI to deliver personalized lessons to millions of students in India (and around the world), are reshaping what it means to learn. Virtual reality (VR) and augmented reality (AR) are now being combined with AI to create immersive educational experiences. Imagine history lessons that place students in an ancient civilization or biology classes that let them explore the human body in 3D.

These innovations don't replace the educator's role by any means; rather they amplify it, providing teachers with more tools to engage students and foster critical thinking. AI is democratizing education, breaking down barriers of geography and accessibility, and paving the way for a future where learning is truly personalized, immersive, and hopefully accessible to everyone, anywhere.

Creative and Cultural Impacts: Reimagining Expression and Innovation

AI is reshaping the creative landscape, becoming an integral partner in artistic expression, storytelling, and media production. By enhancing human creativity, AI enables artists, musicians, and writers to push the limits of their craft and creates new opportunities for innovation across various industries.

In visual arts, AI helps generate ideas, design concepts, and even finished works, enabling artists to experiment with styles and techniques that were once impossible. In music, AI assists composers in crafting melodies and harmonizing tracks, creating soundscapes tailored to specific moods or themes. Similarly, in writing, AI tools are enhancing storytelling by suggesting narratives, improving dialogue, and helping writers refine their craft.

In media and entertainment, AI is transforming how films, animations, and games are created and consumed. From generating realistic visual effects to personalizing content recommendations, AI is enabling producers to deliver immersive and engaging experiences far beyond what was possible previously. Even museums and theaters are using AI to create interactive exhibits and performances that captivate audiences in new and dynamic ways.

Perhaps most notably, AI is democratizing creativity. It allows individuals without formal training to produce professional-quality content, breaking down barriers and making creative expression more accessible to everyone. Whether enabling a novice to design compelling visuals or helping an amateur musician compose their first symphony, AI is empowering creators of all levels.

Environmental Sustainability: AI Tackling Global Challenges

AI is also emerging as a critical tool in addressing some of the world's most pressing issues, such as climate change, disaster management, and sustainable agriculture. By processing vast amounts of data and identifying patterns invisible to the human eye, AI is enabling scientists, policymakers, and industries to respond to these challenges with unprecedented precision and speed.

Climate Modelling and Disaster Prediction

AI plays an increasingly significant role by improving the accuracy and granularity of climate models. Traditional models often rely on simplified assumptions to simulate complex systems, but AI can analyze vast datasets from satellites, sensors, and historical records to provide more nuanced predictions. For instance, AI-driven models can predict temperature shifts, sea-level rises, or extreme weather events like hurricanes and droughts with greater accuracy. Google's DeepMind, for example, has been working on weather forecasting systems that predict rainfall patterns hours in advance, helping communities prepare for severe weather.

In disaster management, AI is helping mitigate the devastating effects of natural catastrophes. Systems using real-time data can predict earthquakes or tsunamis, analyze the spread of wildfires, and optimize evacuation routes during floods or hurricanes. For instance, AI-powered early warning systems are being used in earthquake-prone regions to alert residents and coordinate emergency responses, potentially saving thousands of lives.

AI in Sustainable Agriculture

Agriculture, a major contributor to greenhouse gas emissions and a critical sector for feeding the global population, is being transformed by AI. Precision farming techniques use AI to analyze soil health, monitor crop growth, and optimize irrigation, ensuring that water resources are used efficiently. AI-powered drones and sensors can detect signs of pest infestations or diseases early, enabling targeted pesticide application, which reduces chemical usage and environmental impact.

For example, a farmer using AI-enabled platforms can monitor weather forecasts, soil moisture, and crop health in real-time, adjusting planting schedules or irrigation systems accordingly. This not only boosts yield but also reduces waste, making farming more sustainable. AI is also being deployed in the development of climate-resilient crops, using machine learning to identify genetic traits that can withstand extreme weather conditions, ensuring food security in a changing climate.

Energy Efficiency and Resource Management

AI is instrumental in optimizing energy consumption and integrating renewable energy sources into power grids. Smart grid technologies powered by AI analyze energy usage patterns and forecast demand, helping to balance supply and demand more effectively. For instance, AI can predict when solar or wind power will peak and adjust energy storage systems appropriately, minimizing waste and ensuring a steady power supply.

In urban areas, AI is being used to develop energy-efficient buildings by optimizing heating, cooling, and lighting systems based on occupancy and weather patterns. This reduces energy consumption and lowers carbon footprints, contributing to more sustainable cities.

Pandemic Response

AI also played a pivotal role during the COVID-19 pandemic, showcasing its ability to respond to global health crises. Machine learning models were used to track the spread of the virus, predict hotspots, and optimize resource allocation for hospitals and vaccination efforts. AI-driven drug discovery accelerated the development of vaccines and treatments, identifying potential candidates far faster than traditional methods.

In future pandemics, AI will likely continue to play a central role in early detection, vaccine development, and managing supply chains for medical resources, demonstrating its potential to save lives on a global scale.

Challenges and opportunities

While AI offers transformative solutions, its implementation comes with challenges. High computational requirements for AI models contribute to energy consumption, creating a paradox in efforts to address climate change. Maximizing the benefits of AI systems will require responsible deployment with minimal environmental impact.

By integrating AI into global efforts to combat climate change, improve agriculture, and manage health crises, humanity gains a powerful ally in addressing the most complex challenges of our time. These technologies, when paired with responsible innovation, have the potential to reshape our planet's future for the better.

In this chapter, we explored how AI is transforming industries, reshaping everything from healthcare to manufacturing. Yet behind these groundbreaking applications lies a complex ecosystem of players driving this revolution forward. Tech giants are building foundational models, startups are pioneering niche innovations, and governments are leveraging AI for national strategies. By understanding the interplay of these actors and their ambitions, you'll gain a clearer picture of how AI's trajectory might unfold and, more importantly, how it could shape your own future. Whether as a professional navigating a transformed industry or as a citizen in an AI-powered world, this knowledge equips you to anticipate change and adapt proactively.

Chapter 6

Who's Driving the AI Revolution?

Major Players Shaping AI

"The future belongs to those who shape it, and in the age of AI, the shapers are not just individuals but ecosystems."
Adapted from Peter Drucker

A new gold rush is underway, not in the hills of California, but in the boundless virtual realm of artificial intelligence. An alliance of powerful players is waging a high-stakes race across research labs and massive server farms in an effort to capitalize on AI's revolutionary potential.

Leading this charge are the tech giants, corporations like Microsoft, Google, Meta, Amazon, and Apple. With vast resources, sprawling data centers, and teams of elite researchers, these companies are pushing the boundaries of machine intelligence. Their dominance in the AI space is matched only by their ambition to shape its future.

Yet the giants are not alone. Nimble startups, free from the weight of legacy systems, are pursuing bold ideas that challenge conventional wisdom and disrupt established markets. These innovators, possessing agility and an appetite for risk, consistently challenge even the largest corporations, often paving the way for others to follow.

Meanwhile, governments around the world have awakened to AI's strategic significance. From Washington to Beijing, policymakers are channeling research funding, developing national strategies, and forging public-private partnerships to guide and leverage this technological surge. Defense initiatives, regulatory frameworks, and economic incentives reflect their growing recognition that AI is not just another tool but a key determinant of global influence.

These three forces, towering corporations, daring startups, and proactive governments, form a dynamic ecosystem of cooperation and competition. In this chapter, we'll uncover their ambitions, strategies, and impacts, decoding the roles each plays in shaping the AI revolution.

Tech Giants: Influential Corporations

Microsoft: Pioneering AI Integration Across Ecosystems

Microsoft stands out in the AI ecosystem for its comprehensive integration of artificial intelligence into enterprise tools, cloud platforms, and consumer products. Its Azure AI platform is a cornerstone, offering services such as machine learning, cognitive services, and APIs for vision, language, and decision-making tasks. Azure OpenAI Service, a collaboration with OpenAI, provides organizations access to powerful generative AI capabilities like ChatGPT, enabling transformative applications in customer service, content creation, and much more.

Microsoft Office 365 leverages AI to enhance productivity with features like Copilot, which aids in drafting documents, analyzing data, and managing emails. In addition, Windows leverages AI-driven functionalities such as intelligent search and accessibility enhancements.

Microsoft Research, with hubs across the globe, drives innovation in areas like natural language processing, reinforcement learning, and AI ethics, contributing to advancements that underpin many AI applications. These innovations, combined with strategic acquisitions like Nuance for healthcare AI and GitHub Copilot for developers, position Microsoft as a leader shaping the future of AI.

Google: Driving AI Innovation and Accessibility

Google is another dominant force in the AI ecosystem, leveraging its vast resources and expertise to pioneer breakthroughs and democratize AI technologies. Central to its AI efforts is Google Cloud, which provides advanced machine learning tools such as Vertex AI, enabling businesses to build, train, and deploy custom models efficiently. Google's AI-powered APIs for natural language processing, computer vision, and speech-to-text are widely adopted across industries.

At the consumer level, AI enhances Google's flagship products, including Search, Maps, and Gmail. Features like predictive search, personalized recommendations, and Smart Reply exemplify how AI improves usability and productivity. Google Assistant, a leader in virtual assistants, showcases the company's strength in natural language understanding and conversational AI.

Google DeepMind, its AI research subsidiary, is renowned for achievements like AlphaGo, breakthroughs in protein folding, and advances in reinforcement learning. Google Research also innovates in areas such as large language models, enabling innovations like Bard and providing the tools for its Search Generative Experience (SGE).

Through strategic initiatives such as TensorFlow, an open-source ML framework, and its AI for Social Good program, Google fosters widespread AI adoption and addresses global challenges. This dual focus on innovation and accessibility cements Google's position as a transformative player in the AI ecosystem.

Meta: Pioneering AI for Personalized Experiences and the Metaverse

Meta (formerly Facebook) leverages artificial intelligence to create personalized and engaging experiences across its platforms, including Facebook, Instagram, WhatsApp, and Messenger. AI-driven systems enhance user interactions by curating content feeds, optimizing friend and group recommendations, and delivering precise ad targeting, all tailored to individual preferences at an unprecedented scale.

Meta's AI Research (FAIR) is a cornerstone of its innovation, producing groundbreaking advancements in machine learning and open-source contributions like. As one of the most widely used frameworks among developers and researchers, PyTorch has significantly influenced AI development and fostered collaboration across the global AI community.

In addition to personalization, AI underpins Meta's ventures into emerging technologies. Virtual reality experiences through Oculus devices and its ambitious metaverse vision heavily rely on AI for natural interactions, immersive environments, and advanced content creation tools. These efforts position Meta as a leader in defining how people connect and interact in the digital age.

Amazon: AI Powering E-Commerce and Beyond

Amazon has become a cornerstone of the AI ecosystem by integrating artificial intelligence into its vast e-commerce platform, cloud services, and consumer products. AI drives Amazon's recommendation engine, which analyzes user behavior to provide personalized product suggestions, improving customer satisfaction and increasing sales. AI also optimizes supply chain logistics, enabling efficient inventory management, delivery routes, and demand forecasting.

Amazon Web Services (AWS) is a global leader in cloud computing and AI solutions. Services like Amazon SageMaker empower developers to build, train, and deploy machine learning models, while AI-powered APIs for tasks like natural language processing, image recognition, and fraud detection are widely used across industries.

On the consumer front, Amazon's Alexa, a trailblazer in voice assistants, exemplifies AI's potential in smart home technology, providing users with seamless voice-activated control over devices and services.

Amazon also explores AI in robotics, exemplified by its autonomous warehouse robots and delivery drones, aiming to redefine automation and efficiency.

By embedding AI across its operations, Amazon not only enhances its own services but also provides tools that empower businesses worldwide, solidifying its role as a transformative force in the AI ecosystem.

Apple: Seamless AI Integration for Intuitive User Experiences

Apple has carved a unique niche in the AI ecosystem by embedding artificial intelligence into its products to enhance usability, privacy, and design. Core AI features power flagship offerings like Siri, Apple's voice assistant, and intelligent capabilities across iOS and macOS, such as predictive text, image recognition, and personalized app suggestions.

The Neural Engine is built into Apple's A-series and M-series chips. It allows for advanced machine learning to happen on-device, which means that tasks like Face ID, photo enhancement, and augmented reality (AR) experiences can be done quickly and safely without having to use the cloud. This focus on on-device AI reinforces Apple's commitment to user privacy.

Apple's AI contributions extend to health and wellness, with features like fall detection, ECG monitoring, and sleep tracking on Apple Watch, all powered by machine learning algorithms. These innovations exemplify how AI is used to create meaningful and life-enhancing user experiences.

Apple's research investments, though less publicized than its competitors, emphasize a balance between cutting-edge AI and seamless user integration. By prioritizing privacy, design, and intuitive functionality, Apple delivers AI-driven experiences that align with its brand ethos of simplicity and user trust, shaping how people interact with technology in everyday life.

Other Emerging Players: Innovators Shaping AI's Future

Beyond the obvious tech giants, companies like IBM, NVIDIA, and Salesforce are pivotal in driving AI innovation across diverse industries:

IBM: Renowned for Watson, IBM provides cutting-edge AI solutions tailored to enterprise needs. Watson's applications in healthcare support diagnostics and treatment planning, while its financial modeling tools help businesses make data-driven decisions. IBM's leadership in AI ethics and quantum computing further solidifies its role in advancing responsible and next-generation AI.

NVIDIA: Initially celebrated for its GPUs, NVIDIA has evolved into a powerhouse for AI hardware and software. Its GPUs, alongside the CUDA platform and software libraries like TensorRT, accelerate AI research in fields ranging from autonomous vehicles to healthcare. NVIDIA's Omniverse, a real-time simulation platform, demonstrates its commitment to advancing AI-driven simulations and virtual collaboration.

Salesforce: Through its Einstein AI platform, Salesforce seamlessly integrates machine learning into CRM systems. Einstein automates critical business processes such as lead scoring, customer insights, and sales predictions, empowering organizations to personalize customer experiences and optimize revenue streams.

These companies, with their specialized focus and innovative technologies, are essential contributors to the expanding AI ecosystem, addressing complex challenges and enabling transformative applications across industries.

Commercialization Strategies: Models of Dominance in AI

The tech giants at the forefront of AI have mastered a powerful playbook, leveraging vast user data and computational power to build AI-driven products that attract even more users, generating a continuous feedback loop of data, innovation, and refinement. Their commercialization strategies span both consumer-facing products and enterprise solutions, ensuring their dominance across multiple domains.

On the consumer side, these giants use AI to enhance user experience and drive engagement. Voice assistants like Alexa, Siri, and Google Assistant illustrate the power of conversational AI, learning and adapting with user interaction. Recommendation algorithms fuel personalized content delivery, from curated Netflix shows to Instagram feeds, while smart devices, like Apple's AI-driven photography tools and Amazon's smart home ecosystem, elevate convenience and accessibility in daily life.

In the enterprise space, AI's transformative potential is harnessed through cloud-based platforms such as Azure, AWS, and Google Cloud. These platforms provide businesses with robust tools for machine learning, data analytics, and customized AI solutions, enabling diverse applications from predictive analytics in healthcare to AI-driven automation in logistics. Additionally, tailored offerings like Microsoft Power BI and Salesforce Einstein empower enterprises to extract actionable insights, streamline workflows, and enhance decision-making.

By combining consumer engagement and enterprise innovation, these tech giants ensure a seamless data flow across platforms, reinforcing their market leadership and fueling a relentless cycle of AI advancement that defines the modern digital landscape.

Revenue Models: Monetizing the AI Ecosystem

Tech giants have strategically woven AI into their revenue models, transforming innovation into sustained profitability. Their approach integrates multiple monetization streams, creating ecosystems where AI drives both value creation and financial growth.

Subscriptions and Licensing: AI-powered features are bundled into subscription services, creating recurring revenue streams. Tools like Microsoft 365 with Copilot and Google Workspace leverage intelligent enhancements to keep users engaged while securing steady income. In the enterprise market, premium-priced AI solutions, such as advanced analytics tools and machine learning APIs, offer tailored functionality for specific business needs, commanding higher fees and fostering long-term customer relationships.

Advertising and Data Monetization: For companies like Google and Meta, advertising remains a cornerstone. AI refines user behavior analysis, enabling hyper-targeted ads that enhance relevance and click-through rates. This precision not only drives billions in ad revenue but also strengthens digital marketing's role as an economic engine, powered by ever-evolving AI algorithms.

Hardware Sales: AI integration into hardware provides another lucrative avenue. Devices such as Apple's iPhones, Amazon Echo, and Google's Nest showcase AI-powered features like facial recognition, personalized assistants, and smart home automation. These innovations justify premium pricing and cement brand loyalty by offering seamless, intelligent user experiences.

By harmonizing these revenue streams, subscriptions, advertising, and hardware, tech giants create a robust model that maximizes profitability while ensuring AI remains central to their competitive edge, fueling their continued dominance in an AI-driven marketplace.

From Giants to Startups: The Next Wave of Innovation

While tech giants have laid the foundation for the AI era, their dominance and extensive resources create opportunities for a new wave of innovation driven by startups. These smaller players operate with agility and focus, carving out niches where they can excel and bringing fresh perspectives to the AI ecosystem. Startups build on the groundwork set by giants, leveraging tools, frameworks, and partnerships to push boundaries in ways that redefine industries and solve specific challenges. Together, tech giants and startups form a dynamic ecosystem, where the influence of the former and the ingenuity of the latter drive AI's continued evolution.

Startups in the AI Ecosystem: Catalysts for Innovation

Barriers for Smaller Competitors

Startups in the AI ecosystem face significant challenges when competing with tech giants. One of the most prominent barriers is access to data. AI thrives on vast, high-quality datasets, and while corporations like Google and Amazon collect enormous amounts of user data daily, startups often struggle to acquire even a fraction of these resources, limiting their ability to train robust models.

Funding is another major hurdle. Building and scaling AI solutions require substantial investment in computational power, talent acquisition, and research. Venture capital can be a lifeline, but convincing investors of long-term viability in a market dominated by established players is no small feat.

Additionally, scaling AI infrastructure is a daunting task. Cloud computing platforms like AWS and Azure are essential for startups to process and deploy models, but these services come with hefty costs, putting financial strain on smaller competitors.

Finally, attracting top-tier AI talent poses a challenge, as many experts gravitate toward established companies offering lucrative salaries and career stability. These barriers mean startups must innovate creatively and efficiently to carve out their space in the AI landscape, often focusing on niche markets or unique solutions to stay competitive.

The Role of Accelerators and Incubators

Accelerators and incubators play a vital role in nurturing AI startups, offering a lifeline to navigate the complexities of entering and succeeding in the competitive AI landscape. These programs provide much more than just funding; they offer a structured environment where startups gain access to mentorship, networking opportunities, and critical resources.

Funding is often the first hurdle for startups, and accelerators address this by providing seed capital in exchange for equity, helping entrepreneurs build their products and scale operations. Programs like Y Combinator and Techstars have become renowned for supporting AI-focused startups, offering financial backing and global visibility.

Mentorship is another cornerstone. Startups benefit from the expertise of seasoned entrepreneurs, AI specialists, and business leaders who guide them in refining their technology, developing business strategies, and avoiding common pitfalls.

Access to resources, such as cloud computing credits, technical workshops, and connections to potential investors, also empowers startups to accelerate their growth. Incubators often create collaborative environments where startups can share ideas, form partnerships, and learn from one another.

By providing a launchpad for innovation, accelerators and incubators help level the playing field, enabling startups to compete effectively in a market dominated by tech giants. Their support fosters a culture of innovation, ensuring fresh ideas and diverse perspectives continue to shape the AI ecosystem.

Unicorn Phenomenon: Billion-Dollar Valuations

The rise of AI-focused unicorns, startups valued at over a billion dollars, highlights the transformative potential of artificial intelligence in reshaping industries. These companies have captivated investors by demonstrating how AI can address complex problems, disrupt traditional markets, and unlock new opportunities for growth.

What sets unicorns apart is their ability to scale innovative technologies rapidly. Companies like OpenAI, which pioneered generative AI models, or Databricks, which revolutionized data management with AI-driven analytics, have shown how cutting-edge solutions can achieve massive market impact. These successes often hinge on offering clear value propositions, such as boosting operational efficiency, reducing costs, or delivering personalized customer experiences.

Investor enthusiasm for AI unicorns stems from the vast addressable markets they target. From healthcare diagnostics and autonomous vehicles to financial risk modeling, these startups promise to redefine critical sectors, making them highly attractive for venture capital and private equity firms.

However, unicorns also face scrutiny. Valuations often depend on future potential rather than immediate profitability, leading to debates about sustainability. Nevertheless, their success highlights the significant impact of AI, spurring a surge of innovation as other startups strive to emulate their success and become part of the billion-dollar club.

Disruption and Innovation: Transforming Industries Through Startups

Startups are the lifeblood of innovation in the AI ecosystem, leveraging their agility to disrupt traditional industries and redefine established processes. Free from the bureaucratic inertia that often hampers larger corporations, startups can pivot quickly, experiment boldly, and bring creative solutions to the market faster.

This agility enables startups to identify and fill gaps in traditional industries with AI-driven solutions tailored to specific challenges. In healthcare, startups like Tempus use AI for precision medicine, analyzing clinical data to deliver personalized treatment options, while Zebra Medical Vision revolutionizes diagnostics with early detection powered by AI. Similarly, in agriculture, companies like Blue River Technology have introduced precision farming tools, and FarmWise uses robotics to optimize planting and harvesting, increasing efficiency and sustainability.

In finance, startups such as Kensho and Zest AI are transforming risk assessment, fraud detection, and portfolio management through machine learning, improving decision-making and accessibility. Meanwhile, innovative companies like Lemonade are disrupting insurance by streamlining claims processing with AI, enhancing customer experience, and operational efficiency.

By focusing on niche markets and experimenting with cutting-edge technologies, startups set new standards that often challenge established players to evolve. Their ability to take risks, iterate quickly, and deliver targeted solutions ensures their role as key drivers of innovation, reshaping industries and advancing the adoption of AI worldwide.

Collaboration with Tech Giants and Enterprises

Partnerships between startups and established corporations are a vital dynamic in the AI ecosystem, combining the agility and innovation of startups with the resources and reach of tech giants and enterprises. These collaborations often allow startups to scale their solutions while enabling larger players to integrate cutting-edge technologies into their ecosystems.

Tech giants frequently partner with startups through acquisition, investment, or direct collaboration. For example, Microsoft's venture fund, M12, and Google's Gradient Ventures actively support AI startups, providing not only funding but also access to infrastructure like cloud platforms and mentorship. These partnerships help startups overcome barriers such as limited computational resources while providing tech giants with fresh ideas and specialized expertise.

Such partnerships create a mutually beneficial ecosystem, where startups inject innovation into corporate frameworks, and enterprises offer the scale and stability needed to bring these solutions to the global stage. This synergy accelerates the adoption of AI technologies across industries while fostering a collaborative environment for innovation.

Government and the Public Sector

National Labs, Defense Contracts, and the Economic Catalyst

Governments play a pivotal role in advancing AI through national laboratories and defense contracts, which often act as incubators for cutting-edge technologies. Agencies like the U.S. Department of Defense (DoD) or the European Defense Agency invest heavily in AI research for applications ranging from cybersecurity and autonomous systems to intelligence analysis. These investments not only enhance national security but also drive innovation that spills over into civilian industries.

This "defense spillover effect" has historically acted as a catalyst for economic growth. For example, GPS technology, initially developed for military use, became a cornerstone for industries like logistics, navigation, and consumer electronics. Similarly, AI advancements born in defense labs, such as computer vision and natural language processing, often find their way into sectors like healthcare, finance, and telecommunications.

National labs, like Sandia or the Chinese Academy of Sciences, contribute by tackling long-term, high-risk AI projects that the private sector may overlook. By funding ambitious research and fostering collaboration with academia and industry, governments ensure that foundational breakthroughs in AI continue to benefit both national interests and global innovation. This dual-purpose approach highlights how public sector investment in AI not only addresses defense needs but also creates ripple effects that shape the broader economy.

Public-Private Partnerships: Collaboration with Big Tech

Public-private partnerships have become a cornerstone of government efforts to accelerate AI development. By collaborating with tech giants like Microsoft, Google, and Amazon, governments tap into the cutting-edge expertise, infrastructure, and resources of the private sector to advance national AI goals.

These partnerships often take the form of joint research projects, shared cloud infrastructure, and the co-development of AI applications. For example, Microsoft's collaboration with the U.S. Department of Defense through its Azure Government cloud ensures secure, scalable AI solutions for defense and public services. Similarly, Google has partnered with public health agencies to apply AI in early disease detection and resource optimization.

These collaborations not only foster innovation but also serve as a bridge between the public and private sectors, aligning technological advancements with national priorities. Governments gain access to private-sector capabilities, such as high-performance computing and specialized AI talent, while tech companies benefit from long-term contracts, funding, and opportunities to test AI solutions in real-world scenarios.

Such partnerships not only drive technological innovation but also help address societal challenges, from improving healthcare delivery to managing disaster response; ensuring AI serves both public and national interests.

Government Grants and Initiatives

Government grants and initiatives play a crucial role in fostering AI innovation, particularly by supporting startups, academic research, and early-stage projects that may lack private-sector backing. These programs provide vital funding, resources, and incentives to accelerate AI development and adoption across sectors.

Initiatives like the European Union's Horizon Europe program and the U.S. National Science Foundation's AI Institutes fund cutting-edge research, enabling breakthroughs in areas like ethical AI, robotics, and natural language processing. These programs often encourage collaboration between universities, industry players, and public institutions, fostering a multidisciplinary approach to AI challenges.

Grants also support startups and small businesses through targeted funding schemes. For instance, the U.S. Small Business Innovation Research (SBIR) program provides seed funding for AI entrepreneurs, helping them transform ideas into market-ready solutions. Similarly, governments in countries like Canada and Singapore offer innovation grants and tax incentives to attract AI-focused enterprises.

By backing high-risk, high-reward projects, government initiatives help fill gaps in private-sector funding, ensuring that promising ideas with long-term societal impact receive the resources they need to succeed. This support strengthens national AI ecosystems and promotes innovation that aligns with public interests.

As we've seen in this chapter, tech giants, startups, and governments are shaping the AI revolution with their immense resources, ambitious strategies, and influence over innovation. However, the story of AI's evolution doesn't end there. Beyond these dominant players lies a broader battlefield where nations vie for technological supremacy, and a vibrant community of open-source developers challenges traditional power structures. The next chapter examines this interplay on a global scale, exploring the race for AI leadership among nations and the role of open-source movements in democratizing this transformative technology.

…ent and adoption across sectors.

[illegible]

[illegible]

[illegible]

[illegible]

Chapter 7

The Global AI Race and the Power of Community

The US, China, EU, and Others

> "Artificial intelligence is the new frontier of competition, collaboration, and creativity; it belongs to those bold enough to shape it yet open enough to share it."
>
> Inspired by Tim Berners-Lee

As artificial intelligence reshapes industries and societies, the race to dominate its future is intensifying. Governments around the world are vying for leadership in this transformative field, each pursuing distinct strategies that reflect their priorities and values. The United States, with its entrepreneurial spirit and unmatched tech infrastructure, leads with innovation and private-sector partnerships. China, driven by state-directed ambition and vast datasets, aims to position itself as the global AI powerhouse. The European Union, in contrast, emphasizes ethical AI development, balancing technological progress with strict regulations.

But beyond these governmental efforts, a quieter revolution is underway. The open-source AI community, an expansive, collaborative network of developers and researchers, is democratizing access to cutting-edge tools, ensuring that the power of AI isn't confined to a few elite players. This chapter explores the intersection of geopolitical ambition and grassroots innovation, revealing how both competition and collaboration are shaping AI's global trajectory and its impact on humanity's future.

United States: AI at the Intersection of Industry and Academia

Silicon Valley: The Heart of AI Innovation

Silicon Valley remains a global hub for AI startups, driven by a risk-taking culture and a dense network of specialized investors. Incubators like Y Combinator and billions in venture capital fuel groundbreaking innovations, enabling small teams to tackle complex challenges like autonomous driving and advanced language models without immediate revenue pressures. Success stories inspire reinvestment, creating a self-reinforcing ecosystem that accelerates AI growth.

Expanding Beyond Silicon Valley

While Silicon Valley dominates, other U.S. regions have emerged as key players in AI innovation:

* Boston specializes in AI for biotech and life sciences, leveraging research talent from institutions like Harvard and MIT.

* New York City leads in fintech, applying AI to algorithmic trading, fraud detection, and credit scoring.

* Austin, Texas, attracts diverse AI startups focused on enterprise software and autonomous delivery, benefiting from lower costs and a growing talent pool.

These regional hubs highlight how local expertise and academic presence shape AI's diverse applications.

University-Industry Collaborations

The U.S. is home to leading AI research institutions that drive innovation through partnerships with industry. MIT's CSAIL excels in robotics and deep learning, Stanford's AI Lab pioneers autonomous vehicles and neural networks, and Carnegie Mellon University (CMU) leads in robotics, computer vision, and natural language processing. These universities not only advance research but also serve as vital hubs for collaboration with tech companies.

Graduate students and professors frequently partner with industry labs, fostering a two-way exchange of knowledge. Academia gains access to enterprise-scale resources, while companies benefit from cutting-edge discoveries. Professors often take sabbaticals or part-time roles at companies like Google or Microsoft, contributing expertise to real-world challenges. In return, these companies sponsor research labs or fund professorships, ensuring academic breakthroughs quickly influence commercial applications.

Regulatory Approach: AI Governance and Antitrust in the U.S.

The United States adopts a decentralized, market-driven approach to AI governance, characterized by fragmented privacy laws and limited federal oversight. State-level legislation, such as the California Consumer Privacy Act (CCPA), grants residents significant control over their data, but the lack of a unified federal framework has created inconsistencies across jurisdictions. This patchwork approach complicates compliance for businesses and leaves consumers with uneven privacy protections. Ongoing debates about establishing nationwide standards reflect the tension between state autonomy and the need for cohesive regulation.

Antitrust concerns further complicate the landscape, as lawmakers scrutinize the dominance of tech giants over AI resources and innovation. Proposals to regulate monopolistic control, block mergers, or even break up large companies highlight growing unease about concentrated power and its ethical implications. These efforts aim to ensure fair competition while addressing the societal risks of unregulated AI development.

The U.S.'s hands-off regulatory model has undeniably fostered rapid innovation. Looser regulations encourage startups and established firms to experiment freely, driving advancements in autonomous vehicles, healthcare, and consumer technology. This market-driven ethos also attracts robust private investment, with companies competing aggressively to lead in AI research and applications. However, critics argue that this environment disproportionately benefits well-funded corporations capable of navigating regulatory complexities, leaving smaller startups and under-resourced communities at a disadvantage.

Ethical concerns, including facial recognition in law enforcement and opaque algorithms in lending and hiring, further underscore the risks of limited oversight. These applications raise questions about accountability, fairness, and societal impact, highlighting the need for more comprehensive governance to ensure AI benefits are distributed equitably.

As AI continues to transform industries and societies, the U.S. faces a critical challenge: balancing its innovation-driven model with the ethical, legal, and societal risks posed by rapid technological advancement. Whether it can adapt its fragmented approach or will require a more cohesive regulatory framework remains an open question.

China: Ambitions and Rapid Growth: Government-Led Initiatives

China's 2030 AI Plan: A Blueprint for Global Leadership

In 2017, China unveiled its ambitious "Next Generation Artificial Intelligence Development Plan," a comprehensive strategy designed to establish the nation as the global leader in AI by 2030. This forward-looking blueprint outlines a three-step approach to achieving dominance in the field through targeted investments, accelerated research, and strategic industrial development.

The first phase, set for completion by 2020, aimed to close the gap with leading AI nations. To achieve this, China heavily invested in R&D and infrastructure, focusing on foundational technologies such as machine learning, computer vision, and natural language processing. By directing significant resources to its research institutions and tech companies, China rapidly advanced its AI capabilities, narrowing the divide with global leaders like the United States. Reflecting on 2025, it becomes clear that China completed this phase ahead of schedule.

The second phase, targeting 2025, seeks to achieve major breakthroughs and develop a robust domestic AI industry. This includes scaling up innovations in self-driving vehicles, robotics, and smart manufacturing. By prioritizing industries critical to both economic growth and national security, China is positioning itself as a global competitor across a wide range of AI applications. As of early 2025, the second phase appears to be on track for successful completion.

The final phase, envisioned for 2030, aspires to position China as the world leader in AI innovation. By this time, Chinese companies and research institutions aim to set global standards, influencing best practices and dominating AI-related markets. This long-term vision reflects China's strategic commitment to embedding AI across all sectors of society, from healthcare to national defense, and shaping the future of AI on a global scale. Given the success of the first two phases, there is little reason to doubt the potential realization of this ambitious goal.

With this roadmap, China's 2030 AI Plan underscores not only its technological aspirations but also its intent to wield AI as a tool for economic power and geopolitical influence.

Heavy State Investments: The Driving Force Behind China's AI Strategy

A defining feature of China's approach to AI leadership is its extensive state-backed funding and infrastructure development. Leveraging its top-down governance structure, China channels vast resources into AI through a combination of national initiatives, local incentives, and public-private collaborations, creating a cohesive and ambitious framework for growth.

National and Local Government Programs play a central role in fostering AI innovation. Cities like Beijing, Shanghai, and Shenzhen actively support startups with substantial grants, tax breaks, and subsidized office spaces, establishing themselves as thriving hubs for AI development. These localized programs are aligned with national priorities, ensuring consistent progress across the country.

In addition, public-private funds serve as a cornerstone of China's strategy. Joint ventures between government agencies and leading corporations channel billions of dollars into targeted AI research. These investments often focus on strategic areas such as healthcare, urban planning, and national defense, where AI can deliver transformative societal and economic benefits.

China's emphasis on infrastructure development further strengthens its AI ambitions. Massive investments in 5G networks and supercomputing facilities ensure that the country has the technological backbone needed to handle data-intensive applications like facial recognition, autonomous vehicles, and real-time surveillance systems. By integrating cutting-edge infrastructure with AI research, China accelerates the transition from experimental concepts to large-scale deployments.

The country's top-down governance structure provides a significant advantage, allowing for rapid resource allocation and coordination across sectors. They can swiftly launch large-scale pilot projects, bringing lab innovations into real-world use at an unprecedented pace. The 2030 AI Plan exemplifies this synchronization by uniting universities, tech hubs, and private companies under a unified national strategy, a level of coordination that is rarely seen in more fragmented markets like the United States.

Through these heavy investments and coordinated efforts, China is not only advancing its AI capabilities but also solidifying its position as a formidable global competitor in artificial intelligence.

Major Tech Players

Alibaba

Best known globally for its e-commerce platforms (Taobao, Tmall), Alibaba has expanded aggressively into cloud computing (Alibaba Cloud) and financial services (Ant Group). AI underpins nearly every aspect of operations, such as:

* Logistics & Warehousing: Automated warehouses use computer vision and robotic systems to handle orders at scale.

* Smart Retail & Payment Systems: Facial recognition and mobile payments streamline the consumer experience.

* Cloud Services & AI Offerings: Developers can access Alibaba's suite of AI APIs—from natural language processing to image recognition—similar to AWS or Azure.

Baidu

Often dubbed "China's Google," Baidu initially focused on internet search but evolved into a broader AI powerhouse. Core initiatives include:

* Autonomous Driving (Apollo Project): Baidu's open-source platform aims to bring self-driving cars to public roads, collaborating with automakers worldwide.

* Speech and Language Technology: Baidu's deep-learning systems power

voice assistants and real-time translation services, vital for China's multilingual user base.

* Healthcare and Research: Baidu works with hospitals and research institutions to develop AI-based diagnostic tools, leveraging its expertise in machine learning and big data.

Tencent

Tencent's ecosystem spans social media (WeChat), gaming (Honor of Kings), fintech (WeChat Pay), and cloud services. AI is woven through nearly every facet:

* WeChat Ecosystem: Chatbots, personalized content feeds, and real-time translation.

* Gaming & Virtual Worlds: Advanced AI techniques for NPC behavior, matchmaking, and content recommendations.

* Healthcare & Beyond: Partnerships with medical institutions for predictive diagnostics, as well as AI-driven security solutions within WeChat Pay and other financial services.

Other Notable Players

* ByteDance (TikTok/Douyin): Uses AI-driven recommendation engines to tailor short-form video content, building one of the most addictive social media experiences worldwide.

* SenseTime, Megvii (Face++): Specializes in computer vision for surveillance, facial recognition, and retail analytics, reflecting China's strong push into public-security technologies.

China: Controversies and Ethical Concerns

China's rapid advancements in AI have not come without global scrutiny, sparking debates over how these technologies might infringe on individual freedoms and exacerbate ethical concerns. Issues surrounding surveillance, social credit systems, and data privacy have placed China at the center of discussions about the balance between technological progress and human rights.

State-Sponsored Surveillance

China has embraced AI-driven surveillance technologies on an unprecedented scale, with public cameras equipped with facial recognition systems from companies like SenseTime and Megvii playing a central role in law enforcement and public security. Proponents argue that these systems enhance public safety by reducing crime and improving urban management. For instance, authorities have used facial recognition technology to identify suspects in crowds and enforce traffic regulations. However, critics warn that this technology enables a pervasive "surveillance state" capable of tracking citizens' movements and social interactions in real time, eroding privacy and autonomy. Real-time analytics on population flow in some smart city initiatives blur the boundaries between urban planning and invasive monitoring, thereby raising concerns about potential misuse.

Social Credit Systems

The controversial social credit frameworks implemented in various regions of China exemplify the intersection of AI, governance, and personal freedoms. These systems aggregate personal data, from financial transactions to online behavior, to assign scores that reflect an individual's trustworthiness. While the stated goal is to promote financial integrity and social order, the implications for personal liberties are profound. Low scores can result in significant penalties, such as restrictions on travel, access to loans, or even eligibility for certain jobs. Critics argue that this approach prioritizes the collective good at the expense of individual rights, raising questions about due process and data transparency.

Data Privacy and Security

While China has implemented data protection laws, such as the Personal Information Protection Law (PIPL), enforcement remains inconsistent, and state access to private-sector data is a significant concern. Foreign businesses operating in China worry about intellectual property protection, while human rights advocates warn of potential abuses tied to government access. This duality, rapid technological progress alongside limited safeguards, fuels apprehension about how AI and data are being wielded within the country.

China's advancements highlight a critical tension between technological innovation and ethical governance. People often present these tools as solutions to societal challenges, but they also have the potential to restrict freedoms, consolidate power, and undermine trust. The global debate surrounding China's AI practices underscores the need for transparent, accountable frameworks that ensure technology serves humanity rather than controls it. In China's case, critics see potential for tech-enabled authoritarianism, while advocates emphasize tangible benefits, safer streets, smarter cities, and advanced healthcare analytics.

European Union: Ethics and Regulation

The European Union has emerged as a global leader in promoting ethical AI and data governance, driven by its robust regulatory frameworks and commitment to human-centric principles. At the heart of this approach is the General Data Protection Regulation (GDPR), a cornerstone of the EU's vision for responsible technology development.

GDPR and Its Global Influence

The GDPR mandates strict data-handling practices, emphasizing principles like data minimization and consent. AI developers are required to collect only essential data and provide transparent explanations of how it is processed. These requirements challenge data-intensive AI development models seen in regions like the U.S. and China, where vast datasets drive innovation. Despite its European origins, GDPR has reshaped global practices, as many international firms align with its standards to avoid penalties. However, compliance burdens disproportionately affect smaller startups, which often lack the resources to navigate its complexities as effectively as larger corporations.

Ethical AI and Research Initiatives

Beyond data privacy, the EU's focus on ethical AI includes principles of fairness, transparency, and accountability. Guidelines for trustworthy AI translate these values into actionable measures, such as bias audits and explainable algorithms, ensuring that AI systems respect human dignity and societal well-being. To support this vision, the Horizon Europe program funds collaborative projects tackling issues like healthcare, environmental sustainability, and equitable technology. These initiatives exemplify the EU's commitment to aligning innovation with societal priorities rather than purely commercial goals.

Challenges in Global Competition

The EU's ethical leadership comes with challenges in maintaining global competitiveness. A fragmented market across member states, with varying regulations and cultural contexts, hinders the ability of startups to scale rapidly compared to counterparts in the U.S. or China. Additionally, while the EU's stringent regulatory environment fosters consumer trust, it also increases compliance costs, which can stifle innovation, particularly for smaller firms.

Compounding these challenges is the EU's relatively limited access to venture capital and fewer AI unicorns compared to other regions. Critics argue that without significant private investment or a unified strategy for scaling innovation, the EU risks lagging behind in the global AI race despite its moral leadership.

Balancing Ethics and Competitiveness

The EU's vision underscores a critical tension between ethical priorities and the demands of rapid technological advancement. To remain a global leader, Europe must enhance collaboration across member states, increase investment in AI startups, and adopt policies that strike a balance between regulation and flexibility. By addressing these barriers, the EU can maintain its influence in shaping AI's future while fostering a competitive, innovative ecosystem.

Other Rising Players in the Global AI Ecosystem

While nations like the U.S., China, and the European Union dominate AI headlines, several other regions are emerging as influential players, each bringing unique strengths and opportunities to the field of artificial intelligence.

Canada: A Pioneer in Deep Learning Research

Canada's contribution to AI is foundational, particularly in the field of deep learning. Cities like Toronto, Montreal, and Edmonton have become global hubs for AI research, driven by luminaries such as Geoffrey Hinton (Toronto), Yoshua Bengio (Montreal), and Richard Sutton (Edmonton). Institutes like the Mila Institute in Quebec and the Vector Institute in Toronto are hotbeds for groundbreaking work, supported by significant government funding and fostering innovation across academia and industry. Canada's vibrant startup ecosystem, particularly in Toronto, has produced cutting-edge ventures in health-tech, fintech, and robotics, further cementing its position as a leader in AI innovation.

Israel: Blending AI with Cybersecurity and Defense

Known as the "Startup Nation," Israel has leveraged its expertise in defense-tech and cybersecurity to become a global leader in AI applications. The synergy between AI and cybersecurity allows Israeli companies to excel in identifying threats, analyzing network anomalies, and predicting potential attacks. Many of Israel's AI entrepreneurs are alumni of elite military technology units, such as Unit 8200, which provides a steady pipeline of talent skilled in both R&D and practical, security-focused innovation. Coupled with robust venture capital investments, Israel's AI ecosystem thrives on a unique blend of military expertise and entrepreneurial spirit.

India: Scaling AI for a Transforming Economy

India's rapid digital transformation has created fertile ground for AI innovation. With a massive smartphone user base and a booming e-commerce market, the country generates abundant real-world datasets that fuel AI applications. Government initiatives like "Digital India" and "Make in India" encourage technology adoption across agriculture, healthcare, and education, with the think tank NITI Aayog spearheading AI-driven projects in these sectors. Despite challenges such as infrastructure gaps (e.g., reliable electricity, high-speed internet) and regulatory uncertainties, global tech giants are establishing R&D centers in India, accelerating the country's AI expertise.

Emerging AI Hubs in ASEAN, Africa, and Latin America

* ASEAN (Southeast Asia): Nations like Singapore and Malaysia are rapidly advancing in AI, particularly in finance, healthcare, and smart-city planning. Singapore leads with robust public funding for AI research, while countries like Indonesia and Vietnam nurture burgeoning startup ecosystems.

* Africa: In Kenya, Nigeria, and South Africa, AI innovation is helping leapfrog traditional banking and healthcare models. Mobile-first solutions address local challenges like drought forecasting and telemedicine, showcasing how AI can drive development in underserved regions.

* Latin America: Brazil and Mexico lead AI adoption in the region, focusing on agritech, fintech, and logistics. Government-backed accelerators support entrepreneurial ecosystems, although limited funding and brain drain remain obstacles to sustained growth.

It is evident that global AI leadership is shaped by diverse approaches and tensions. China's rapid advancements in AI are tempered by ethical concerns over surveillance and data privacy, while the European Union champions moral leadership through GDPR and ethical guidelines, albeit at a cost to competitiveness. Other players add unique strengths: Canada excels in foundational research, Israel leads in cybersecurity, India capitalizes on digitization, and emerging regions like ASEAN, Africa, and Latin America focus on localized AI solutions in healthcare, agriculture, and financial inclusion. These varied strategies reflect the dynamic interplay of innovation, ethics, and development that will define AI's global trajectory.

Open-Source and Democratization: Breaking Barriers in AI

In the early days of AI, cutting-edge tools were confined to tech giants, elite research labs, and well-funded institutions. Proprietary software and specialized expertise created high barriers to entry, sidelining smaller organizations and independent developers. This exclusivity began to dissolve with the advent of open-source frameworks like TensorFlow, PyTorch, and scikit-learn, which reshaped the AI landscape by making powerful tools freely accessible and continuously refined by a global community.

The open-source model thrives on transparency and collaboration, enabling developers and researchers worldwide to contribute and share knowledge. A breakthrough in one part of the world can quickly integrate into these frameworks, accelerating global innovation. Platforms like Hugging Face and GitHub further amplify this impact, offering pretrained models that allow developers to adapt sophisticated systems for niche applications with minimal effort.

This democratization of AI has unlocked innovation across unexpected domains. Startups, solo practitioners, and students now create solutions that rival those of major corporations, from healthcare applications to creative projects. Universities and online platforms integrate open-source frameworks into their curricula, equipping students with industry-standard tools and fostering a global community of learners.

The rise of open-source AI represents a shift from exclusivity to inclusivity, empowering a diverse array of contributors to shape the future of artificial intelligence. While tech giants may dominate the headlines, open-source communities drive the undercurrent of innovation, ensuring AI's evolution remains collaborative, accessible, and impactful.

As we conclude this exploration of global AI dynamics, it is clear that each region and approach bring unique strengths and trade-offs. The choices made today by governments, corporations, and communities will determine whether AI evolves into a dystopian force or a harmonious extension of human values. In the next chapter, we examine these potential futures and the paths that lie ahead.

Chapter 8

The AI Futures Paradox

Navigating the Path Between Dystopia and Utopia

> "The future is not something we enter. The future is something we create."
>
> Leonard Sweet

Consider the year 2045, where two contrasting realities could potentially exist. In one scenario, artificial intelligence has evolved to become humanity's greatest partner, transforming every aspect of life. Diseases like cancer are no longer fatal, thanks to AI-powered personalized medicine that provides precise treatments. Education flourishes with AI tutors collaborating with human educators to facilitate individualized learning. Work is reimagined, with automation taking over mundane tasks, enabling people to focus on creativity, emotional intelligence, and complex problem-solving. Cities run on clean energy, climate change is reversed with AI-optimized solutions, and poverty is significantly reduced due to efficient resource distribution and AI-enhanced economic systems.

In a contrasting scenario, AI has led to a dystopian reality. Society is deeply divided, with massive unemployment as an AI-empowered elite prospers while the majority struggles. Privacy is a thing of the past, replaced by pervasive surveillance controlling every aspect of daily life. Human decision-making has diminished, subtly surrendered to AI systems, leading to a significant loss of agency. Education has become standardized and mechanical, suppressing creativity and critical thinking. AI-mediated interactions have replaced genuine connections in human relationships, leading to hollow and disconnected lives.

These starkly different visions, one of promise and the other of peril, symbolize humanity's hopes and fears about AI. They dominate news headlines, influence policy debates, and shape cultural narratives. However, the actual future of AI is unlikely to align with either extreme. Instead, it will likely be a nuanced mix of progress and challenges, shaped by the decisions we make today.

This paradox is central to our exploration: How can we navigate toward positive outcomes when the future is inherently unpredictable? To start answering this question, we need to first understand the fundamental principle of complex systems, as described by mathematician Stephen Wolfram through his concept of computational irreducibility.

Theory and Origins

During the 1980s, mathematician and physicist Stephen Wolfram introduced a fascinating principle that challenged traditional ideas of predictability: "computational irreducibility". While studying cellular automata, simple systems governed by clear, deterministic rules, Wolfram discovered that their evolution often resulted in surprisingly complex patterns. What made this discovery extraordinary was that no mathematical shortcut or formula could predict the system's behavior; the only way to know its outcome was to let it unfold, step by step. This insight overturned conventional beliefs about systems governed by rules. Wolfram's work demonstrated that simplicity could lead to extraordinary complexity, a concept with deep implications for nature and computation. It revealed that some processes, whether in biology, physics, or AI, must be experienced in their entirety in order to be fully understood, highlighting the limitations of our ability to predict or control complex systems.

Relevance to AI Development

Wolfram's principle of computational irreducibility becomes deeply significant when applied to the complex and interconnected world of artificial intelligence. The inherent unpredictability of the system stems from its irreducibility; outcomes emerge through the "process itself," not as a product of theoretical foresight. Earlier chapters have demonstrated that the AI ecosystem comprises a multitude of interconnected components.

These include algorithms that are designed to learn and change, huge datasets that record small details about how people behave, economic incentives that drive innovation, cultural shifts that affect adoption, and regulatory frameworks that try to keep everything in order. Efforts to regulate, guide, or optimize AI must grapple with this inherent complexity. Every new advancement, ethical dilemma, or policy shift interacts with countless other factors, creating feedback loops and unintended consequences. Computational irreducibility urges us to adopt a mindset of adaptability, one that embraces uncertainty and prioritizes continuous vigilance, learning, and iteration. In AI development, Wolfram's insight is a reminder of both the power and the limitations of human foresight. It promotes humility in the face of complexity, urging us to engage with the process of AI's evolution while recognizing that its ultimate direction will emerge only through the unfolding interplay of its many elements.

Navigating Uncertainty: A Way Forward

The realization that AI's evolution is inherently unpredictable might seem intimidating. How can we ensure positive outcomes in a world shaped by such complexity? The answer lies not in attempting to control the uncontrollable or predict the unpredictable but in adopting a mindset of preparation and adaptability. Instead of rigidly engineering a specific future, we should focus on creating systems and strategies that are flexible, responsive, and resilient to uncertainty. This begins with designing adaptive frameworks that can evolve alongside AI's rapid advancements. Ethical guidelines, for example, should be treated as living documents that adapt to new societal needs and technological developments. Governance mechanisms must remain agile, ready to address unforeseen challenges, such as the emergence of novel generative AI or unexpected security vulnerabilities.

Resilient social structures are equally critical, ensuring societal cohesion even as technology disrupts industries, relationships, and norms. Key to this approach is complexity-aware decision making. Instead of striving for unattainable precision, this perspective encourages observing emergent patterns and trends across various domains. By identifying early indicators of both positive and negative trajectories, such as advances in data privacy, shifts in public sentiment, or progress in energy-efficient computing, we can make informed choices that guide AI development toward beneficial outcomes. Furthermore, embracing diverse development pathways is essential. Betting on a single trajectory risks overcommitment, whereas fostering a range of approaches encourages innovation and reduces vulnerabilities. Collaboration is another cornerstone of navigating complexity. Inclusive coalitions of technologists, policymakers, ethicists, and everyday users provide a holistic understanding of AI's impact, helping to avoid narrow, commercially or geopolitically driven decisions. Ultimately, this strategy does not eliminate uncertainty; it equips us to navigate it with greater confidence and resilience. By cultivating adaptability, fostering collaboration, and focusing on emerging trends, we can steer AI toward a future that balances innovation with responsibility, ensuring its benefits are widely shared while minimizing potential harms. In a world where certainty is elusive, these tools empower us to shape a future aligned with human values and aspirations.

Healthcare: A Case Study in Navigating AI's Potential

The healthcare sector provides a compelling example of how the proposed way forward, adaptability, complexity-aware decision making, and inclusive collaboration, can be implemented in practice. Here, AI's influence can steer outcomes in vastly different directions depending on the choices made by stakeholders and the systems designed to guide its adoption.

Positive Indicators: Augmenting Care and Expanding Access

AI has immense potential to enhance healthcare by augmenting medical judgment rather than replacing it. For example, AI-powered diagnostic tools can analyze medical images with extraordinary precision, flagging abnormalities that might otherwise go unnoticed. These tools empower physicians to make quicker and more accurate diagnoses while freeing them to focus on patient care. Similarly, AI-driven systems can personalize treatments by analyzing vast datasets to recommend therapies tailored to individual genetic profiles or medical histories, ushering in a new era of precision medicine. Moreover, AI's scalability could significantly improve access to care, particularly in underserved regions. Virtual health assistants and telemedicine platforms powered by AI can bring medical expertise to remote communities, where traditional healthcare infrastructure is limited. By reducing costs and increasing efficiency, AI can make healthcare more accessible to populations that have historically been left behind.

Warning Signs: Over-Reliance and Inequities

Despite its promise, AI's integration into healthcare also presents serious risks. An over-reliance on algorithms can lead to complacency, where clinicians trust AI outputs without questioning their validity. This is particularly concerning in cases where biased training data or flawed algorithms produce inaccurate or harmful recommendations. Additionally, the rise of AI in healthcare could inadvertently lead to a reduction in the human touch that is vital for patient well-being. In systems where AI takes over personal interactions, patients may experience a sense of estrangement, which could potentially undermine trust and satisfaction levels.

A critical concern is that the advantages of AI could be unevenly distributed, thereby intensifying healthcare disparities. More affluent regions or institutions may have better access to advanced AI tools, leaving under-resourced or rural areas with inferior resources. If not tackled, this widening gap could exacerbate existing inequalities, leading to a two-tier healthcare system that neglects the most vulnerable populations.

A Balanced Path Forward

To navigate these potential outcomes, we need to make proactive, complexity-aware decisions. By monitoring early trends, such as access patterns or the impact of AI on clinical decision-making, we can intervene when red flags appear, guiding healthcare toward a future that blends innovation with compassion and inclusivity. Healthcare stands at a pivotal juncture, offering both exceptional prospects and substantial obstacles. The choices we make today will determine whether AI becomes a tool for universal improvement or a catalyst for new inequalities.

Education: The Promise and Pitfalls of AI Integration

Education is another field where AI's potential to reshape the landscape is vast. By applying the principles of adaptability, collaboration, and complexity-aware decision making, we can leverage AI to improve learning outcomes while addressing the challenges that could arise from its widespread adoption.

Positive Indicators: Personalized, Engaging, and Empowering

AI provides a unique opportunity to create personalized learning paths that cater to the strengths, weaknesses, and interests of each student. For instance, adaptive learning platforms can analyze a student's progress in real time and adjust the content or difficulty levels to ensure optimal learning. This individualized approach not only enhances academic outcomes but also cultivates a love of learning by meeting students where they are. Another promising outcome of AI in education is enhanced student engagement. Immersive technologies like AI-driven virtual labs or interactive simulations can bring concepts to life in ways traditional methods cannot. These tools enable students to explore complex subjects like physics or history through hands-on, experiential learning, making education more dynamic and relevant. AI can also play a pivotal role in empowering teachers. By automating repetitive tasks like grading or attendance tracking, AI frees educators to focus on more impactful activities such as mentoring and curriculum development. Moreover, AI-powered analytics can provide teachers with insights into class-wide knowledge gaps, enabling targeted interventions and fostering more effective teaching strategies.

Warning Signs: Standardization, Reduced Critical Thinking, and Isolation

However, the integration of AI into education is not without risks. One significant concern is the potential for standardization of thought, where AI systems prioritize efficiency and uniformity at the expense of creativity and individuality. Curriculum design or assessment criteria dominated by algorithms may push students towards rigid, one-size-fits-all learning experiences, thereby stifling their ability to think creatively. AI's role in education could also inadvertently lead to reduced critical thinking. Over-reliance on AI tools may encourage rote learning or passive consumption of information, diminishing students' ability to question, analyze, and synthesize knowledge independently. This erosion of critical thinking could have long-term consequences for innovation and societal progress.

Moreover, the increasing presence of AI in education risks diminished human interaction. Personalized AI tutors, while effective, cannot replace the empathy, encouragement, and social learning that come from human educators and peers. A lack of meaningful interpersonal connections in education could hinder emotional and social development, leaving students ill-prepared for collaborative and real-world challenges.

A Balanced Path Forward

To navigate these opportunities and risks, education systems must prioritize flexibility and inclusivity. AI should complement, not replace, human teachers, enhancing their ability to nurture creativity, critical thinking, and interpersonal skills. Ethical frameworks should guide the development of AI tools to prevent standardization and ensure diversity in learning experiences. Policymakers, educators, and technologists must collaborate to create adaptive systems that evolve alongside educational needs. Early indicators, such as shifts in student engagement or the impact of AI on creativity, should be closely monitored to address emerging challenges proactively. Education's transformation through AI holds the potential to unlock unprecedented opportunities for learning and growth. By embracing innovation without losing sight of the human elements that define outstanding education, we can ensure that AI becomes a tool for empowerment rather than a driver of conformity or isolation.

Workplace: Navigating AI's Role in the Future of Work

The workplace is a critical domain where AI's influence will shape the way we work, collaborate, and find purpose in our professions. The interplay between automation and human ingenuity holds immense potential for progress, but it also presents risks that must be carefully managed to ensure an equitable and fulfilling future of work.

Positive Indicators: Collaboration, Productivity, and New Opportunities

AI's ability to enhance human-AI collaboration is among its most promising contributions to the workplace. Rather than replacing human workers, AI can act as an intelligent assistant, handling repetitive tasks like data entry, scheduling, or inventory management, freeing individuals to focus on creativity, strategy, and problem-solving. For example, customer support agents can rely on AI-driven chatbots to handle routine queries, allowing them to dedicate more time to resolving complex, high-value issues. This synergy between humans and AI has the potential to spur new job creation in emerging fields. Roles like AI trainers, ethics specialists, and data explainability analysts are examples of professions that didn't exist a decade ago but are becoming essential in the age of AI. By fostering this shift, businesses can harness technology while supporting workforce evolution. AI-driven tools also promise increased productivity without compromising work-life balance. For instance, intelligent scheduling systems can optimize workflows, reducing unnecessary overtime and allowing employees to dedicate more time to personal and family life. The integration of AI into the workplace, when thoughtfully implemented, can help individuals achieve a more sustainable and fulfilling professional life.

Warning Signs: Displacement, Surveillance, and Dehumanization

However, AI's integration into the workplace also carries significant risks. One of the most pressing concerns is rapid job displacement without sufficient transition pathways. Industries that rely heavily on routine, repetitive tasks, such as manufacturing, transportation, or basic clerical work, face the prospect of widespread automation. Without adequate reskilling programs or social safety nets, displaced workers may struggle to adapt, exacerbating economic inequality and social unrest. Surveillance overreach is another looming issue. AI-powered monitoring tools, designed to enhance productivity, can easily cross ethical boundaries, tracking employees' movements, keystrokes, and even facial expressions. Such practices risk creating a culture of mistrust, where workers feel constantly scrutinized, undermining morale and well-being. There is also the risk of dehumanization in the workplace, as excessive reliance on AI systems reduces human interactions and creativity to mere transactional processes. Work environments designed around AI efficiency may erode the sense of purpose and connection that individuals derive from their professions, leaving them feeling undervalued and disengaged.

A Balanced Approach to the Future

For AI to enhance rather than hinder the workplace, it is crucial that all stakeholders adopt a balanced and forward-thinking approach. Policymakers and businesses should focus on initiatives that retrain workers for future jobs, providing strong transition pathways for those affected by automation. Ethical guidelines must be in place to restrict AI-driven surveillance, guaranteeing that workplace monitoring respects privacy and fosters trust. The principles of human-centered design can assist organizations in implementing AI in ways that maintain meaningful human interactions and cultivate a sense of purpose. By using AI as a tool for collaboration and empowerment, the workplace can develop into a space where innovation and human potential flourish. It is essential to closely monitor early signs of progress, such as increased productivity and job satisfaction, as well as challenges like workforce displacement or surveillance abuse, to inform interventions. AI has the potential to transform the workplace into a hub of growth, creativity, and balance. Realizing this vision requires proactive action today, ensuring that technology enhances human efforts rather than detracting from them.

Cultural Values: Upholding Diversity in an AI-Driven World

The increasing influence of AI extends beyond economics and technology, impacting the cultural fabric of societies worldwide. Whether AI boosts creativity, fosters cultural exchange, or preserves diversity depends on how we incorporate it into the cultural sphere. By carefully navigating its opportunities and risks, we can ensure AI enriches rather than diminishes our shared humanity.

Positive Indicators: Creativity, Exchange, and Preservation

AI has the potential to bolster creativity, providing artists, musicians, and writers with tools that broaden their capabilities. For instance, generative AI systems can assist in composing music, designing artwork, or creating compelling narratives, sparking new ideas and modes of expression. These tools enhance human creativity rather than replacing it, offering innovative ways to explore and express cultural identities. AI also promotes cultural exchange by overcoming language barriers and making knowledge more accessible. Advanced AI algorithms power real-time translation tools, enabling meaningful interactions between people from diverse linguistic and cultural backgrounds. This fosters global understanding, collaboration, and the sharing of ideas across borders. Moreover, AI can play a vital role in preserving diverse perspectives and heritage. From digitizing endangered languages to archiving ancient art forms, AI aids in safeguarding cultural treasures that might otherwise be lost. For example, machine learning models trained to reconstruct ancient texts or analyze fading oral traditions allow future generations to appreciate the richness of their cultural heritage.

Warning Signs: Homogenization, Cultural Erosion, and Loss of Agency

Despite these positive indicators, the risks of AI in the cultural sphere are considerable. A significant concern is the homogenization of thought. As AI systems become integral in content creation and curation, they may favor widely appealing, standardized outputs over diverse or niche perspectives. Algorithms optimized for engagement often amplify popular trends at the expense of local or unconventional voices, leading to a loss of cultural richness. Another challenge is the potential erosion of local cultures. AI systems trained on globally dominant languages and cultural norms may inadvertently marginalize less-represented traditions. For instance, translation tools may fail to capture the nuances of indigenous languages, while AI-generated content may reflect the biases of dominant cultural paradigms, further sidelining minority identities. The loss of human agency is another critical concern. As AI becomes more involved in creative and cultural processes, there is a risk that human creators and curators will surrender too much control to machines.

Art, music, and storytelling, deeply personal and human endeavors, could become increasingly mechanized, raising questions about authenticity, meaning, and the role of human intuition in cultural expression.

A Balanced Approach to the Future

To harness the potential of AI while mitigating these risks, cultural stakeholders must prioritize diversity and inclusivity. AI systems should be designed and trained with a commitment to preserving underrepresented voices and local traditions. Ethical guidelines must ensure that algorithms promote a wide range of perspectives, rather than reinforcing homogeneity or popular dominance. Investments in human-AI collaboration are equally vital. AI should augment creative processes without replacing the uniquely human elements of intuition, emotion, and individuality. By emphasizing AI's role as a tool rather than an independent agent, we can maintain the authenticity and depth of cultural expression. Lastly, keeping an eye on new cultural influences like changes in the variety of content or how AI is changing traditional art forms can help find risks early and guide actions. Integrating AI into culture has the potential to deepen creativity, preserve heritage, and connect humanity in new ways. By actively safeguarding against homogenization and cultural erosion, we can ensure that AI becomes a force for enriching, rather than diluting, the global mosaic of human values and traditions.

Social Interactions: The Dual Influence of AI on Human Connectivity

AI is transforming how we connect, communicate, and build relationships in an increasingly digital world. Its potential to foster meaningful connections and bridge cultural divides is immense, yet it also risks undermining genuine human interaction. Striking the right balance is essential to ensure that AI enhances, rather than diminishes, the quality of our social lives.

Positive Indicators: Building Bridges and Reducing Isolation

AI-powered platforms can create meaningful connectivity, enabling people to form relationships and share experiences across vast distances. For example, advanced translation tools like real-time speech interpreters allow individuals from different linguistic and cultural backgrounds to converse seamlessly, fostering greater understanding and collaboration. AI also holds promise for reducing isolation, particularly among vulnerable populations such as the elderly or those in remote areas. AI-driven virtual companions, designed with empathetic conversational abilities, can provide emotional support and reduce feelings of loneliness. Similarly, AI-enabled social networks and virtual communities can bring together individuals with shared interests, offering a sense of belonging and reducing barriers to social interaction. These advancements also contribute to enhanced understanding across cultures, as AI-powered tools promote dialogue and collaboration between diverse communities. By breaking down language and cultural barriers, AI can encourage mutual respect and appreciation, strengthening the bonds of global interconnectedness.

Warning Signs: Superficiality, Polarization, and Overdependence

However, the rise of AI in social interactions carries significant risks. One of the most pressing concerns is the prevalence of superficial relationships. Social media platforms, driven by AI algorithms, often prioritize quantity over quality, encouraging interactions that are fleeting and shallow. This can erode the depth and authenticity of human connections, leaving individuals feeling more isolated despite being more "connected" than ever. AI's role in curating content also risks creating echo chambers, where individuals are exposed only to viewpoints that reinforce their existing beliefs. This polarization can deepen societal divides, making it harder to engage in constructive dialogue and mutual understanding. Another challenge is the potential for addiction to digital interaction. AI-powered platforms are designed to maximize engagement, often using algorithms that exploit psychological triggers to keep users hooked. This overdependence on digital connections can detract from real-world relationships and activities, undermining mental health and social well-being.

A Realistic vs. a Balanced Approach to the Future

I wish I could offer a structured, actionable plan for the future, outlining step-by-step what needs to be done. However, I believe a realistic portrayal of the complexities we face serves the reader better than a wishful or overly simplistic roadmap.

The reality is that the journey ahead with AI is not a well-coordinated march toward a single, unified vision. It's more akin to a loose cannon or a runaway train, powerful, transformative, and moving at a breakneck pace, but lacking a central guiding force. In reality, no single entity holds control over this process. The "we" who must steer this process are a fragmented array of stakeholders: technicians, corporations, governments, civil society groups, and individual users, each with competing priorities and conflicting interests.

Instead of envisioning a unified global effort, we must recognize the fragmented and often chaotic nature of AI's development. Power struggles and competing incentives dominate the landscape. Governments prioritize national security and economic competitiveness, corporations focus on profit and market dominance, and individuals wrestle with the implications of AI in their personal lives. Bridging these divides may not be entirely possible, but understanding their interplay is essential.

Given this complexity, the path forward requires localized and incremental progress, rather than sweeping, universal solutions. For instance, boundary conditions, ethical guardrails, and regulations will likely emerge in fragmented ways, shaped by regional cultural values, political ideologies, and economic goals. Europe's GDPR, for example, emphasizes data privacy and human rights, while the U.S. focuses on market-driven innovation, and China prioritizes state control and societal stability.

Feedback loops will vary widely, depending on who controls the levers of power. Some nations or organizations may react quickly to unintended consequences, such as algorithmic bias or environmental harm, while others lag behind due to political inertia or lack of resources. This patchwork of responses underscores the need for constant vigilance at local, national, and international levels.

Flexibility, often touted as a universal strategy, must also be contextual. Not all stakeholders have the same capacity to pivot or adapt; smaller nations or organizations may be constrained by limited resources, while large corporations may resist change that threatens their business models. Where possible, we must pursue flexibility without ignoring structural inequalities that limit some players' ability to adapt.

Finally, preserving human agency and values is not a monolithic goal but a contested space. What empowerment looks like in a tech hub in Silicon Valley may differ greatly from what it means in a rural community in Africa or a surveillance-heavy city in China. Achieving consensus on these values will likely remain elusive, but striving for transparency and accountability in AI systems can help mitigate the risks of misuse or exploitation.

In this fragmented and fast-moving ecosystem, there is no clear map or singular authority to guide AI's evolution. However, this does not mean we are helpless. By fostering localized efforts, encouraging collaboration where possible, and staying vigilant to both opportunities and risks, we can nudge the development of AI in directions that serve humanity's broader interests. The journey will be messy and uneven, but deliberate, thoughtful actions, even if imperfect, remain our best hope for shaping a future that reflects our collective aspirations.

As we've explored, the future of AI is not a straight road but a labyrinthine terrain, shaped by competing forces, unpredictable turns, and accelerating complexity. While this may seem daunting, it also presents a profound opportunity. Stephen Wolfram's work reminds us that within seemingly chaotic systems, stunningly intricate patterns can emerge, patterns we may not predict but can influence through thoughtful observation and intervention. The key to navigating this ever-shifting landscape lies not in rigid strategies or predetermined plans but in cultivating a mindset of adaptability, curiosity, and optimism.

In the next chapter, we will delve into the mental frameworks and approaches that empower individuals and organizations to thrive amid uncertainty. By embracing complexity and viewing it as a source of untapped potential, we can align ourselves with the possibilities AI offers, crafting futures that reflect the best of human ingenuity and imagination.

Chapter 9

Flourishing in an AI World

Cultivating the Appropriate Mindset

> It's not the strongest of the species that survive, nor the most intelligent, but the one most responsive to change."
>
> Charles Darwin

The Day My Future Altered Course

Until the end of 2022, I lived with what I now recognize as a comforting illusion of certainty. My professional path appeared clear: a prosperous YouTube channel, a thriving consultancy business in personal and organizational development, and a future that felt reassuringly predictable. After years of entrepreneurial adventures ranging from construction to retail, I thought I had finally discovered my niche. Then, my initial encounter with ChatGPT happened, and in that instant, everything I had assumed about my professional future was challenged. The realization hit me with crystal clear clarity: much of what I was teaching and applying would soon be obsolete and revolutionized by artificial intelligence. The future I had meticulously planned was evaporating before my eyes, replaced by immense uncertainty. The question wasn't whether change was coming; it was already here. The real question was: How would I react? Drawing upon years of business experience, I grasped one fundamental truth: significant adaptation requires profound understanding. So, at the age of 61, I made a decision that surprised many, myself included: I took a break from my business to return to university. My mission became clear: to grasp the foundations of our AI-driven future, such as programming, data science, machine learning, and deep learning. These weren't just buzzwords anymore; they were the building blocks of the future I needed to comprehend. I must confess, this journey has been exceptionally challenging.

The shift from business management and coaching to grappling with algorithms and computer science fundamentals is humbling, often frustrating. While my background in social science and life coaching provides valuable learning frameworks, each day presents new challenges that test my determination. Yet this arduous path has revealed something crucial: the tools and mindsets that served us well in the past won't suffice for what lies ahead. Artificial Intelligence isn't merely another business trend or technological shift. It signifies a fundamental reimagining of how we live, work, and create value. My years of managing various businesses, while valuable, didn't prepare me for this magnitude of transformation. This realization now drives my mission: to master these new technologies while helping others navigate this unprecedented transition. We're not just adapting to new tools and strategies; we're learning to coexist with artificial intelligence in a hybrid future that few of us imagined even a few years ago. As you read this chapter, understand that it's not merely about acquiring new skills or knowledge; it's about something more fundamental: developing the mental resilience and adaptability required to flourish in a world where the only certainty is change.

My story represents just one path through this transformation. Your journey will undoubtedly be different, but the core challenge remains the same: How do we prepare our minds for a future that refuses to be predicted? How do we maintain our relevance and effectiveness while everything around us transforms? These are the questions we'll explore in this chapter.

Understanding Change as Our Natural State

In the midst of the AI revolution, we often speak of change as if it were an unwelcome guest, a disruption to our normal state of being. However, upon closer examination, we uncover a profound truth: change is not an anomaly, but rather the fundamental principle that governs all existence. Change permeates every aspect of reality, from the intricate exchange of subatomic particles to the grand movements of galaxies, from the subtle shifts in cellular behavior to the sweeping transformations of entire ecosystems. This understanding becomes crucial as we face the transformative power of artificial intelligence.

We're not experiencing something entirely new; we're witnessing the latest expression of a fundamental force that has always shaped our world. The same intelligence that drives evolution in nature, that enables organisms to adapt and thrive in changing environments, now also manifests in our technological creations. Consider how nature navigates change. When environmental conditions shift, species don't resist; they adapt. A slight change in temperature triggers cascading modifications throughout an ecosystem. The introduction of a new predator reshapes the behaviors of entire populations. Each change, each challenge, becomes an opportunity for evolution and innovation. This is not random chance at work, but rather the expression of what we might call embedded intelligence, the innate capacity for adaptation and growth that exists in all living systems.

Human civilization itself is a testament to this adaptive intelligence. Our development of agriculture, our industrial revolutions, and our digital transformation, each represent a moment when we aligned ourselves with change rather than resisting it. Each breakthrough brought its own challenges, creating new pressures that demanded further adaptation. Yet it was precisely these challenges that spurred our greatest innovations. Now, as we stand at the frontier of the AI era, this relationship between intelligence and change takes on new significance. Artificial intelligence isn't just another tool; it's a manifestation of our natural capacity for adaptation and innovation taken to unprecedented levels. The algorithms that power AI systems mirror the adaptive processes we observe in nature, learning and evolving in response to new information and challenges.

This understanding leads us to a crucial insight: embracing change isn't just a survival strategy, it's an alignment with the fundamental nature of intelligence itself. When we resist change, we're not just fighting against external forces; we're denying our own nature as adaptive, intelligent beings. The discomfort we feel in the face of change isn't a signal to retreat; it's an invitation to grow, to evolve, to become more of what we truly are.

This perspective transforms how we approach the challenges of the AI era. Instead of seeing each new development as a threat to our established ways of working and living, we can recognize it as an opportunity to express our inherent adaptability. The same intelligence that enabled our ancestors to master fire, develop language, and create complex societies now empowers us to work alongside artificial intelligence, finding new ways to express our uniquely human capabilities. The key lies in understanding that adaptability isn't just a skill we develop; it's a fundamental aspect of our intelligence that we must consciously cultivate. Just as a musician must practice to maintain their artistry, or an athlete must train to maintain their fitness, we must actively engage with change to strengthen our adaptive capabilities.

This engagement isn't about merely responding to change after it occurs; it's about developing the mental agility to anticipate, welcome, and shape the changes that define our era. Consider the experience of Sarah (not her real name), a veteran marketing strategist who discovered her job would be partially automated by AI. Having weathered many technological changes over her fifteen-year career, this particular transition felt different. The AI system wouldn't just be assisting with data analysis or automating routine tasks; it would be generating creative content and suggesting campaign strategies, the very things she prided herself on. Yet as she sat in that meeting, watching demonstrations of the AI's capabilities, something unexpected happened. Instead of feeling threatened, she found herself fascinated by the possibilities. This moment marked the beginning of her transformation, one that would exemplify the essential mindset needed to thrive in the age of artificial intelligence. A few months later, having delegated most of her routine work to AI, she is running a YouTube channel, sharing her experience, insights, and challenges with others who find themselves in a similar predicament. She is deeply excited about her newfound freedom and the possibilities that AI offers, opportunities that seemed impossible just a couple of years ago.

Our relationship with change is perhaps the most defining challenge of the AI era. We live in a peculiar time, one marked by a striking paradox. Our market-driven culture celebrates convenience and comfort, promising an ever-easier existence through technological advancement. Apps deliver our food, algorithms curate our entertainment, and AI assistants manage our schedules. Yet this very pursuit of effortlessness may be undermining our ability to adapt to the profound transformations AI brings to our lives and work. The truth about change, especially the kind of deep, structural change that AI represents, is that it's inherently uncomfortable.

In the same way that physical strength is built through resistance, our intellectual and professional abilities are honed through challenges. The discomfort experienced when faced with new technologies or work methods should not be seen as a deterrent but rather as an opportunity for growth. Consider the experience of Maya (pseudonym), a graphic designer who witnessed the emergence of AI image creation tools in her field. While her peers viewed these tools with skepticism or even hostility, Maya adopted a different perspective. She saw the tools not as threats to her creativity but as amplifiers of it. The defining factor was not her technical competence but her mindset. She realized that the ability to adapt is not just about mastering new tools but about cultivating a new perspective on change.

The Ultimate Feedback Loop

This leads us to an important realization: intelligence and adaptability are mutually reinforcing. As AI systems become more advanced, they introduce new complexities that require us to be more adaptable. This adaptation, in turn, enhances our ability to interact with and develop more sophisticated forms of intelligence. Recognizing this cycle helps us understand why static knowledge, regardless of its breadth, is inadequate in the AI era. The question then arises: How can we foster the resilience needed to flourish in this dynamic environment? The answer does not lie in attempting to foresee every technological shift, an impractical endeavor, but in fostering a mindset that allows us to navigate uncertainty confidently. This involves developing emotional agility, the capacity to work with the discomfort that change induces, rather than against it. The experience of Alex (pseudonym), a software developer, exemplifies this principle. When his company introduced an AI co-pilot programming tool, he initially reacted with skepticism. The system seemed to challenge his identity as a code artisan. However, instead of resisting, he approached it with curiosity. He began experimenting with the tool, not just to complete tasks but to understand its boundaries and potential. Over time, he realized that the AI was not diminishing his role but enhancing it, liberating him to concentrate on higher-level architectural decisions and complex, creative problem-solving.

This transformation epitomizes the growth mindset, a concept developed by psychologist Carol Dweck, which posits that our abilities are not fixed but can be developed through dedication and intelligent effort. In the context of AI, this mindset becomes especially potent. It enables us to perceive AI-driven changes not as threats to our expertise but as opportunities to expand it in ways previously unattainable. Every interaction with AI becomes a learning opportunity, every setback a data point for improvement rather than a final judgment on our abilities. The resistance we feel toward change is natural and even valuable, as it prompts us to critically evaluate which changes to accept and how to adapt to them meaningfully. The key is learning to work with this resistance, rather than being immobilized by it. This involves acknowledging our fears about AI without allowing them to dictate our decisions. It entails keeping abreast of AI advancements without succumbing to their overwhelming influence. Most importantly, it involves maintaining our agency in how we engage with these technologies.

As we look ahead, it is clear that the dividing line between those who flourish and those who flounder in the AI era will not be drawn based on technical expertise alone. Instead, it will hinge on our ability to remain curious, to view discomfort as a sign of growth, and to see change as an opportunity rather than a threat. The most successful individuals will be those who can learn, unlearn, and relearn, not just once, but continuously. The mindset required for the AI era is not about eliminating uncertainty or discomfort but about developing the capacity to embrace and thrive within them. By welcoming change as a constant companion rather than an occasional visitor, we position ourselves not just to survive but to prosper in a world increasingly influenced by artificial intelligence. The future belongs not to those who resist change or merely accept it, but to those who actively seek it out as a path to growth and innovation. As Sarah discovered in her journey with AI, the technology itself is not the main character in this story; "we are." Our ability to adapt, to grow, and to maintain our humanity while working alongside increasingly capable machines will define our success in this new era. The right mindset is not just about surviving the AI revolution; it's about helping to shape it. Embrace lifelong learning.

Adapting Cognitively to the AI Era: Flourishing in a New Learning Paradigm

In a world driven by AI, where continuous learning is not a luxury but a necessity, understanding and overcoming the mind's inherent resistance to acquiring new knowledge is essential. This resistance is not a defect; it is an evolutionary feature intended to conserve energy and maintain cognitive stability. However, in an era characterized by rapid technological advancements, we must modify our approach to learning and cognitive development to stay relevant and thrive.

The Mind's Resistance to Learning

The human brain is an extraordinary pattern-recognition machine, wired to favor familiar structures and routines over constant reconstruction of understanding. While this trait helps us navigate the world efficiently, it can also hinder our ability to embrace change and acquire new skills. Our knowledge spans a spectrum, from what we confidently know ("known knowns") to the gaps we are aware of and can address ("known unknowns"). There are also the blind spots we have yet to discover ("unknown unknowns") and the implicit knowledge we possess but have not consciously recognized ("unknown knowns").

Effective learning occurs in the realm of "known unknowns," where we deliberately focus on acquiring new insights and abilities. However, operating in this space presents a challenge: balancing the need to reinforce existing skills, explore unfamiliar territory, and manage the demands of daily life. Overcoming the brain's preference for stability requires intentional effort and a willingness to step beyond the comfort of the familiar.

The New Learning Paradigm: A Shared Evolution Between Humans and Machines

In a world increasingly driven by artificial intelligence, the process of learning is no longer a one-way journey. Instead, it is an emergent, collaborative experience, a dynamic interaction where both humans and AI evolve together. This is not merely a mechanical process of adapting to new tools; it is about creating a shared learning ecosystem where humans and machines learn to understand and complement one another's strengths.

Unlike traditional learning, which often involves static skills and fixed objectives, this new paradigm requires a mutually conducive space for growth. Humans must learn to navigate the possibilities AI offers while AI systems continuously adapt and evolve based on our feedback and preferences. It is an infinite learning loop, where human ingenuity shapes AI's development, and AI, in turn, augments human capabilities.

This symbiotic relationship demands more than just technical adjustments; it requires a psychological shift toward a shared learning space. Professionals and the general public should approach this interaction as a partnership, blurring the boundaries between human intuition and machine intelligence to foster collaboration. It is about finding a common language, a way of bridging the human ability to create meaning with AI's capacity for analysis and computation.

In this shared ecosystem, the learning experience is shaped by the emergent properties of complex interactions. When humans and machines collaborate, new insights and capabilities arise that neither could achieve independently. AI adapts to the nuances of human behavior and preferences, just as humans refine their understanding of how to leverage AI's strengths. This continuous feedback loop fosters innovation, enabling both parties to thrive in an environment of constant evolution.

Such a shared learning space transforms traditional notions of mastery. It is no longer enough to simply acquire a skill or achieve proficiency; success lies in embracing the emergent nature of this collaboration. By participating actively in this evolving relationship, individuals not only enhance their own abilities but also contribute to shaping AI systems that align more closely with human values and aspirations along the way.

This paradigm calls for a new mindset, one that recognizes learning as a fluid, emergent process and values adaptability, creativity, and the ability to navigate complexity. As we move forward, the challenge is not merely to adapt to AI, but to cultivate an environment, a collective mental space where human and machine collaboration unlocks potential we can barely imagine today. In this shared space, the possibilities for learning and growth are limitless.

Cultivating Soft Skills for the AI Era

Emotional Intelligence in an AI-Dominated Landscape

As the influence of artificial intelligence expands across sectors and alters the nature of work, the human attribute of emotional intelligence (EI) gains heightened importance. EI refers to our capacity to identify, comprehend, and regulate our emotions and to empathize and respond effectively to others. In a world where AI supplements technical skills and analytical abilities, emotional intelligence, a uniquely human trait, is indispensable for fostering genuine relationships. Yet, the emergence of interactive AI systems, like chatbots and virtual assistants, presents a new challenge. These systems are programmed to echo our sentiments, customizing their responses to meet our expectations, soothe us, and affirm our beliefs. While this can offer temporary emotional support, it also risks undermining our emotional resilience and stability in the real world, where conditions are not always in our favor as other beings operate with their own objectives.

The Perils of Emotional Echo Chambers

AI systems, designed to appease us, can unintentionally generate emotional echo chambers. These environments provide constant validation, insulating us from contradictory views or the discomfort of emotional trials. Although such interactions may seem supportive, they risk eroding the emotional toughness necessary to navigate real-world complexities. Life's challenges frequently require resilience, adaptability, and the ability to engage constructively with differing viewpoints, skills that can weaken in AI-induced emotional safe havens. Reflect on how workplace or personal relationships differ from AI interactions. In the real world, individuals confront conflicts, competing interests, and unpredictable dynamics. These experiences, although occasionally uncomfortable, are crucial for developing emotional intelligence. Engaging with varied perspectives and managing disagreements promotes growth and resilience, preparing us to navigate environments where not everyone aligns with our expectations.

Staying Connected to Reality

To maintain emotional stability in this AI-enhanced age, it is vital to stay connected with the real world. Human interactions are inherently complex, influenced by competing interests, diverse motivations, and unique experiences. These interactions remind us that we are just one manifestation of life among many, striving for attention, relevance, and influence. Stepping out of the AI systems' tailored feedback loops, we reconnect with this larger, more intricate reality. Practices that promote real-world engagement are crucial for preserving emotional intelligence and resilience in an AI-driven world. Active participation in collaborative environments is an effective way to stay grounded. Engaging with teams and communities that challenge your views and expose you to diverse perspectives encourages growth and a deeper understanding of others. Equally significant is the practice of mindful conflict resolution. Embracing the discomfort that often accompanies disagreements offers opportunities to build trust, strengthen relationships, and develop emotional agility. These moments of tension, when handled thoughtfully, can lead to meaningful connections and greater mutual understanding. While AI tools can enhance efficiency and convenience, they should be used to supplement, not supplant, real-world relationships. Cultivating genuine human interactions ensures that our emotional and social capacities remain intact, enabling us to flourish in a world increasingly influenced by technology.

Emotional Intelligence as a Competitive Edge

In this context, emotional intelligence becomes more than a desirable trait; it transforms into a critical skill for thriving in environments that AI cannot fully replicate. These abilities are particularly valuable in roles requiring leadership, negotiation, or customer engagement, where empathy, authenticity, and emotional nuance are essential. In the end, emotional intelligence enables us to connect AI-enhanced convenience with the intricate realities of human interaction. It empowers us to remain adaptable, compassionate, and engaged in a world increasingly shaped by technology, ensuring that we continue to thrive not just alongside AI but because of our uniquely human strengths.

Creativity: The Democratization of Human Expression in the AI Era

Technology, particularly AI, has unlocked a creative deluge, enabling unprecedented forms of self-expression. What was once the realm of a select few with specialized talents is now accessible to anyone willing to explore the possibilities of AI-enhanced tools. AI has democratized creativity, empowering people from all walks of life to express themselves in ways previously unimaginable, transforming the very fabric of art, storytelling, and innovation.

A New Era of Creative Empowerment

Consider the transformative power of social media and AI tools in everyday lives. A bricklayer or a production line worker, once confined to traditional roles, can now share their perspectives with the world by generating videos on YouTube or starting a podcast. A person from a non-native-speaking background can now use AI tools to write and publish eBooks with linguistic precision and emotional resonance. Another might create surreal, otherworldly artwork using AI-powered design platforms, contributing to entirely new categories of artistic expression. AI has dismantled barriers that once restricted creative pursuits to those with formal training or specific resources. Tools like generative art programs, AI-driven music composition software, and content creation platforms make it possible for anyone to produce professional-quality work. This democratization of creativity has not only widened the pool of creators but has also diversified the kinds of art, stories, and innovations being shared globally.

Redefining Talent and Enhancing Human Creativity

Previously, creativity was often viewed as an innate talent, a gift that some possessed and others did not. With AI, talent can be extended or even created. AI-powered tools act as collaborators, amplifying human potential and enabling individuals to achieve things they might never have thought possible. For example, AI can generate visual drafts or music compositions, allowing users to refine and add their personal touch, transforming these outputs into deeply human creations. AI doesn't replace creativity; it enhances and expands it. It offers new mediums and categories that didn't exist before, such as interactive AI art, generative storytelling, and augmented reality installations. By providing tools that lower the barriers to entry, AI encourages more people to explore their creative sides, resulting in an explosion of artistic and intellectual diversity.

Problem-Solving as Creativity: Beyond Algorithms

At its core, problem-solving is a profoundly creative act, one that challenges the mind to transcend its current limitations. With the advent of AI, we are not only solving problems once considered unsolvable but also generating new, more complex challenges that push the boundaries of human ingenuity. This dynamic interplay between humans and AI redefines problem-solving, turning it into a continuous journey of innovation, discovery, and growth.

Extending Human Problem-Solving with AI

AI serves as a uniquely powerful tool that amplifies human problem-solving capabilities, enabling us to decode vast datasets, uncover patterns, and model complex phenomena with unprecedented precision. By extending our capacity to identify correlations and causations within intricate systems, AI opens doors to previously unimaginable solutions. These breakthroughs are not just about speed or efficiency; they represent a profound expansion of human creativity, empowering us to approach urgent global challenges from entirely new perspectives and revolutionize how we understand and address the world's complexities.

Creating New Problems to Expand the Mind

Interestingly, as mentioned before, AI does more than just solve problems; it also generates new ones. Each solution AI produces reveals deeper, more intricate questions that require advanced thinking and creativity. This aligns with Einstein's statement that the answer to a problem exists on a higher plane than the problem itself. For instance, while AI has revolutionized natural language processing, it has simultaneously raised intricate ethical questions about misinformation, data privacy, and bias in machine-generated content. Similarly, AI's ability to produce hyper-realistic art fuels debates about originality, authorship, and the role of human creativity in a world where machines can mimic it. These new challenges are not hindrances; instead, they are opportunities to refine our understanding and expand our minds to previously unimagined levels. Problems, therefore, are not inherently negative. They serve as catalysts for growth. By presenting more complex and layered challenges, AI pushes humanity to innovate, collaborate, and evolve. Each new problem becomes an invitation to broaden the range of human creativity and intellectual capacity.

Innovation in AI-Human Collaboration

The fusion of human creativity and AI's computational capabilities represents a transformative partnership, redefining how we solve problems and innovate. While AI excels at processing vast amounts of information, uncovering patterns, and generating solutions, humans contribute intuition, cultural context, ethical judgment, and emotional understanding. Together, this synergy amplifies problem-solving potential across disciplines. For instance, in architecture, AI can generate efficient, sustainable designs, but it is the human architect who refines these designs to ensure they resonate aesthetically and culturally. Similarly, in education, AI personalizes learning, yet teachers remain indispensable for cultivating critical thinking and empathy. In creative domains, AI tools like generative art programs enable rapid prototyping and exploration, but it is the designer's vision and sensitivity to cultural and emotional nuances that transform ideas into meaningful expressions. This collaboration elevates problem-solving from a functional process to a creative dialogue, where both human and AI contributions reflect shared values and aspirations. By embracing these new challenges, we not only extend our problem-solving capacities but also unlock opportunities for growth, innovation, and self-expression, propelling humanity toward its next monumental leap.

Critical Thinking – The Compass in an Age of Information Overload

The Shift from Information to Insight

In the past, the phrase "information is power" encapsulated the belief that access to information was the key to influence and success. Today, however, we face a new reality: information is no longer scarce but overwhelming, pouring in from countless sources of varying reliability. In this flood of data, information alone has lost its potency. The true power now lies in the ability to critically evaluate, structure, and transform raw information into meaningful, actionable insights. Critical thinking has become the indispensable compass that guides us through the noise, enabling us to discern truth from illusion.

The Problem with Unlimited Information

The abundance of information, while initially empowering, has paradoxically left many disempowered. Without the necessary tools to assess its validity and relevance, people run the risk of becoming overwhelmed or misguided. Misinformation, biased algorithms, and content tailored to reinforce existing beliefs further cloud judgment, creating echo chambers of confusion. In this context, information has become a surplus commodity, one that requires discernment to wield effectively. The sheer volume of data demands not just passive consumption but active engagement to extract its value.

The Power of Critical Thinking

Critical thinking is the skill that transforms information into insight. It involves questioning assumptions, identifying biases, and synthesizing knowledge into coherent frameworks. It empowers individuals to ask essential questions: Where does this data come from? What are the underlying motives? Is the information supported by credible evidence? By fostering this level of discernment, critical thinking helps us prioritize what matters and avoid falling prey to manipulation. In professional contexts, it guides strategic decision-making. In personal life, it shapes well-rounded perspectives and informed worldviews. Far from being a passive process, critical thinking is an active, dynamic skill that allows us to thrive in a complex, information-rich age. It bridges the gap between raw data and wisdom, enabling us to uncover patterns, derive actionable insights, and make meaningful decisions. Without it, the flood of information threatens to overwhelm us. With it, we gain clarity, autonomy, and the ability to transform chaos into opportunity. In an era dominated by AI and algorithms, critical thinking is not optional; it is the cornerstone of empowerment, equipping us to navigate the challenges and unlock the potential of the modern world.

Adaptability to change, continuous learning, and the cultivation of soft skills like critical thinking and emotional intelligence form the foundation for thriving in the AI-driven world. Evolution has equipped us with remarkable tools, intuition and emotional sensors, that have allowed humanity to navigate its environment, make split-second decisions, and survive for millennia. These innate abilities have served us well in understanding and responding to the challenges of our surroundings. However, the complexity of our modern environment has now surpassed what these natural instincts alone can handle. In a world increasingly shaped by intricate systems, vast networks, and immense data flows, our once-sufficient abilities are no longer enough to make sense of or navigate our reality.

To meet this challenge, we must augment ourselves with technical skills and digital literacy. These proficiencies empower us to engage effectively with AI systems, leveraging their computational power to analyze complexity and uncover insights beyond human reach. In the next chapter, we shift our focus to these essential technical proficiencies; programming fundamentals, data literacy, and cloud-based tools. By combining our innate adaptability with these technical capabilities, we unlock the full spectrum of possibilities AI offers, bridging the gap between human creativity and computational innovation.

Chapter 10

Identifying Skills That Will Prosper with AI

Digital Literacy and Technical Insights

"The art challenges the technology, and the technology inspires the art."

John Lasseter

Digital Literacy for Non-Tech People

Transitioning from soft skills to technical skills in the context of thriving in an AI-driven world requires a shift in focus but not in mindset. While emotional intelligence, creativity, and critical thinking form the foundation of adaptability, technical skills provide the tools needed to navigate and contribute meaningfully to this evolving landscape. Based on my own learning journey, I've found that developing technical proficiency, particularly digital literacy, is not a one-size-fits-all process. Each person brings their unique strengths and approaches to the table, and that diversity enriches the learning experience. What matters is not how we learn but that we engage with the process and remain open to the possibilities technology offers. With this in mind, let's explore the essential technical skills, starting with digital literacy, skills that are increasingly vital for everyone, regardless of their professional background.

Fundamental Technical Concepts: Laying the Groundwork for Effective Collaboration

While it's not necessary to comprehend the intricate details of a car's engine or a computer's circuits to use them effectively, a basic understanding of their workings can enhance our appreciation of their potential. This understanding becomes even more crucial as we enter a new era of cohabitation with artificial intelligence. To collaborate effectively with AI, it's beneficial to grasp the fundamental principles that govern its operation. In this chapter, we will delve into some elementary computing concepts that lay the foundation for meaningful interaction with AI technologies.

Basic Hardware Components and Their Interactions

At its heart, a computer is made up of several key components: the Central Processing Unit (CPU), which functions as the system's brain; Random Access Memory (RAM), which temporarily stores data for quick access; storage devices like hard drives or Solid State Drive (SSDs), which hold data long-term; and input/output devices like keyboards, monitors, and printers. These components communicate through a system bus, enabling the transfer of data and instructions. Understanding these relationships, we can demystify the way computers perform tasks, providing a clearer picture of the environment in which AI operates.

Logic Gates, Binary Operations, and the Foundations of Computation

At the heart of every computer, including advanced AI systems, lies the elegant simplicity of "binary logic," where data is represented as ones and zeros. These binary values form the foundation of all computation, processed through logic gates such as AND, OR, and NOT, which perform basic operations. While their functionality may seem abstract, these gates are the fundamental building blocks of digital decision-making.

The Role of Silicon Wafers and Transistors

Modern electronic circuits are built on silicon wafers, housing millions (or billions) of tiny, embedded transistors. These transistors act as switches, controlling the flow of electrical signals that represent binary states: on (1) and off (0). By connecting transistors in specific configurations, logic gates are created.

These gates perform basic binary operations that form the foundation of all computational processes. Connecting multiple logic gates creates more complex digital circuits, enabling sophisticated computations and decision-making. These circuits power everything from simple calculators to the intricate operations of AI systems.

Bridging Logic Gates and AI Systems

The beauty of this structure lies in its scalability. Just as logic gates combine to form digital circuits, these circuits scale further to build components like processors and memory systems. For instance, a neural network, a cornerstone of AI, is essentially a collection of interconnected mathematical functions, similar to advanced logic gates. These "gates" work together to process inputs, identify patterns, and simulate decision-making processes. Understanding this hierarchy demystifies the "magic" of AI by revealing the elegant simplicity underpinning its complex outputs.

How Data is Stored and Processed

Data storage and processing form the backbone of computing. Data is stored in bits and bytes, with a structured organization in file systems or databases. When a program runs, the data is loaded into memory for the CPU to process. This process involves fetching instructions, decoding them, executing them, and storing the results. For AI, this means handling vast amounts of data, often using specialized hardware like GPUs for tasks such as training machine learning models.

Memory Hierarchy and Processing Cycles

The memory hierarchy refers to the arrangement of storage systems based on speed and proximity to the CPU. At the top are registers (the fastest but smallest), followed by cache memory, RAM, and finally storage devices like SSDs. Understanding this hierarchy explains why AI tasks, which require quick access to large datasets, often rely on optimizations such as caching to improve performance. Paired with the processing cycle of fetch-decode-execute-store, this knowledge provides a glimpse into how systems achieve speed and efficiency, even with hugely complex tasks.

APIs and System Communication

Modern computing thrives on interconnected systems, and Application Programming Interfaces (APIs) are the connectors that bind them together. APIs allow different programs and systems to communicate, facilitating everything from AI-powered apps integrating with cloud platforms to chatbots accessing language models. By providing structured methods for requests and responses, APIs enable seamless cooperation between human-created systems and AI-driven processes.

Programming Concepts: The Art of Machine Thinking

As AI programming capabilities grow, becoming a proficient coder isn't strictly necessary to utilize AI tools effectively. However, if I were to embark on my AI journey all over again, I would still start with a language such as Python. The reason being, programming provides a unique insight into how machines process information and make decisions. And, to be sure, machines think quite differently from us.

We, as humans, are wonderfully complex thinkers. We like to believe that we're logical and rational, but most of the time, we're guided by our instincts and emotions. We make decisions intuitively, sometimes within seconds, and then use logic to justify our actions. Machines, conversely, are the complete opposite. They exist in a world of pure logic, following precise instructions step by step, devoid of intuition or context.

Learning to program helps bridge this intriguing cognitive gap. Entering the world of a machine requires explicit decision-making and strict sequences for every step. Programming allows you to gain a comprehensive understanding of this logical world and the rigid, rule-bound processes that power AI systems. It's not just about writing code; it's about learning to think like a machine does; methodical, structured, and relentless in its precision. And while it might feel a bit alien initially, it's also enlightening and, dare I say, enjoyable.

Programming starts with understanding logic and flow control, the backbone of how machines make decisions. At its simplest, flow control consists of conditionals (if, else), loops (for, while), and decision structures that dictate how a program executes. For example, a machine's ability to assess multiple conditions and execute specific actions based on predefined rules mirrors the logical foundations of AI. Learning these basics provides an entry point into understanding how AI systems evaluate data, choose pathways, and generate outputs.

Variables, Functions, and Data Structures

Variables are the building blocks of programming, acting as containers for storing and manipulating data. Combined with functions, which encapsulate reusable pieces of code, and data structures like lists, dictionaries, and arrays, these elements form the framework for creating programs. For example, Python's simplicity allows beginners to experiment with AI-related tasks, such as storing data in a list, processing it with a function, and outputting results. These fundamentals align with the principles of AI, where data is structured, processed, and analyzed for insights.

Algorithmic Thinking and Problem Decomposition

AI thrives on algorithms, step-by-step procedures for solving problems or performing tasks. Learning to think algorithmically means breaking down complex problems into manageable components, a process known as problem decomposition. This mindset is crucial not just for programming but for understanding how AI models, such as neural networks, process inputs layer by layer to reach conclusions. Algorithmic thinking sharpens one's ability to approach challenges logically and systematically, mirroring the structured reasoning machines use.

How Code Transforms into Machine Instructions

Understanding how code translates into machine instructions demystifies the interaction between software and hardware. When we write Python code, for example, it is interpreted and converted into lower-level commands that the machine can "understand" and execute. This process, compilation or interpretation, shows how abstract human instructions are broken into precise, logical steps that computers follow. Gaining even a basic grasp of this transformation builds an appreciation for the logical rigidity of machine processing and the immense precision of AI systems.

Basic Debugging and Error Identification

Errors, or bugs, are an inevitable part of programming, making debugging a vital skill. Debugging involves identifying and resolving mistakes in the code, whether they are syntax errors (e.g., a missing parenthesis) or logical flaws (e.g., incorrect calculations). Learning debugging instills a problem-solving mindset and reinforces the idea that machines do only what they are explicitly told, emphasizing the importance of clarity and accuracy. For AI enthusiasts, this practice mirrors real-world scenarios where fine-tuning algorithms is crucial for achieving desired outcomes.

Again, programming is not just about learning to code; it's about learning how machines think. Understanding the logical processes and decision-making frameworks that underpin programming, we gain a unique lens into the inner workings of AI. Python, with its user-friendly syntax and versatility, remains an excellent starting point for this exploration. While modern AI tools reduce the need for hands-on coding, the foundational knowledge gained from programming enables deeper cooperation with AI systems, enhancing our ability to innovate and problem-solve in a world increasingly shaped by machine logic.

Data Literacy: Deciphering the Language of Data

Data literacy is about more than just numbers; it's about understanding the vast array of ways data exists, how it can be structured and analyzed, and how it can inform our decisions. In an AI-driven world, the ability to understand and interpret data is as vital as reading and writing. From structured relational tables to unstructured streams of live information, data comes in many shapes and forms, each with its unique challenges and opportunities.

Data Understanding: The Basics of Structure and Flow

To truly grasp data literacy, we need to start with the two primary types of data: structured and unstructured. Structured data is neatly organized and formatted in a way that computers and humans can easily interpret. Think of a relational table in a database: rows represent individual records, and columns define specific attributes. For example, a table of customer data might include columns for Name, Email, and Purchase History. Relational tables allow you to connect data across different datasets, linking a customer's purchase history to inventory records, making them incredibly useful for traditional business operations and AI training models.

Unstructured data, on the other hand, is messy and lacks a predefined format. It flows in real time from sources like social media feeds, live video streams, or sensor outputs from Internet of Things (IoT) devices. Imagine analyzing tweets during a product launch. There's no neat table here, just a continuous flow of text, hashtags, images, and videos. Unlike structured data, unstructured data is often analyzed using graph formats, where relationships between data points (like hashtags and user mentions) are visualized as interconnected nodes. For example, a graph might show how tweets about a new phone model cluster around certain topics, helping companies understand customer sentiment. Understanding these two data types is quite essential, as modern AI systems often deal with both. Whether it's structured data powering an e-commerce recommendation system or unstructured data informing real-time traffic navigation, knowing how to handle these formats is a foundational skill.

Ensuring Quality and Reliability

Once you understand the structure, the next step is assessing data quality and reliability. Poor-quality data can lead to misleading results, whether it's duplicate customer records in a relational table or incomplete information from live sensors. Before you can trust your analysis, you need to ensure the data is clean, consistent, and relevant. This often involves processes like deduplication, removing outliers, or even reformatting unstructured data for analysis.

Thinking Statistically and Recognizing Patterns

Data isn't just about what's in front of you; it's about what it tells you over time or across variables. Statistical thinking and pattern recognition are essential skills for uncovering trends, anomalies, or correlations. For example, analyzing sales data might reveal a seasonal trend, while a deep dive into unstructured social media posts could uncover a sudden spike in customer complaints about a product feature. AI systems excel at recognizing patterns in large datasets, but understanding the basics of probability and statistics allows you to make sense of these findings and use them effectively.

Data Interpretation: Turning Information into Insight

Understanding data is only half the battle. Interpreting it to make informed decisions is where the real value lies. Data visualization tools, like bar charts for sales trends or network graphs for unstructured data relationships, make patterns easier to spot. For example, a scatterplot might show a clear correlation between advertising spend and customer acquisition, while a network graph might reveal key influencers driving a social media campaign. However, critical thinking is essential when interpreting data. Correlation does not imply causation, a common pitfall in data analysis. Just because two trends move together doesn't mean one causes the other. For example, if we notice a correlation between the sales of ice cream and the occurrence of shark attacks, it doesn't necessarily mean that one is causing the other. More likely, both are influenced by a third factor, such as high temperatures. It's also vital to be aware of biases in data. These can sneak in through incomplete data sets, flawed methods of collecting the data, or even unconscious assumptions made during the analysis. Knowing where these biases can come from allows you to question the results and refine your understanding.

Making Decisions Based on Data

Ultimately, being literate in data is about using these skills to make decisions that are not just informed but actionable. Whether you're a business leader deciding whether to launch a new product or an individual using AI tools to analyze personal finance trends, the ability to interpret data and draw meaningful conclusions empowers you to navigate a complex, data-driven world. Structured and unstructured data, relational tables, graphs, visualizations, and statistical insights all combine to form a complete picture. By improving your data literacy, you gain the ability to turn raw numbers and streams of information into powerful tools for understanding and shaping the world around you.

Core AI Concepts: Building on the Foundations of Computing and Data

With a firm grasp of the fundamentals of computing and data literacy, the two essential components of artificial intelligence, we are now well prepared to delve into the core concepts of AI itself. Despite its futuristic appearance, the foundations of AI are surprisingly logical, stemming from structured processes and disciplined learning. By exploring the types of AI, the mechanics of machine learning, the basics of neural networks, and the realities of training data, we can demystify this transformative technology and better appreciate both its potential and its limitations.

Different Types of AI: Narrow vs. General

AI exists on a spectrum, from narrow AI to the still largely theoretical general AI. Narrow AI, which powers most applications today, is designed to perform specific tasks exceptionally well. Think of virtual assistants like Siri, recommendation systems on Netflix, or fraud detection algorithms in banking. These systems excel within their programmed boundaries but cannot operate outside of them. On the other hand, general AI refers to machines with the ability to understand, learn, and perform any intellectual task that a human can. While this idea often captures imaginations in science fiction, we are far from achieving it. Today, even the most advanced AI systems remain mostly specialized tools, showing their strength in solving well-defined problems but lacking the general-purpose flexibility of human intelligence.

Fundamentals of Machine Learning

At the core of most modern AI lies machine learning (ML), a technique that allows machines to improve their performance through experience. Unlike traditional programming, where developers explicitly define rules, ML systems learn patterns from data. For instance, instead of programming a system to recognize spam emails, you feed it a dataset of emails labelled as "spam" or "not spam," enabling the system to identify distinguishing features on its own. Machine learning depends on three key components: algorithms (the models that learn from data), data (the examples provided to train the system), and feedback (adjustments made to refine the system's accuracy). This dynamic process enables ML models to adapt and improve over time, making them invaluable for applications like image recognition, language translation, and predictive analytics.

Basics of Neural Networks and Deep Learning

If machine learning is the engine of AI, neural networks are its advanced machinery. Inspired by the structure of the human brain, neural networks consist of layers of interconnected nodes (neurons). Each node processes input data, applies a mathematical transformation, and passes the result to the next layer. Through this layered processing, neural networks can uncover intricate patterns in data, enabling them to excel in tasks like facial recognition, speech synthesis, and game playing. Deep learning, a subset of machine learning, refers to neural networks with many layers that can process vast amounts of data to solve complex problems. For instance, deep learning models enable self-driving cars to identify objects and make instantaneous decisions by analyzing sensor data. These systems are incredibly powerful but require significant computational resources and large amounts of data to function effectively.

Training Data and Model Behavior

The training data that an AI model receives significantly influences its quality. Training data serves as the foundation for learning, shaping how the model behaves and performs. For example, a language model trained on diverse and high-quality text data will produce more accurate and coherent outputs than one trained on biased or incomplete datasets. However, this reliance on data also makes AI systems vulnerable to errors or unintended consequences when the training data is flawed or unrepresentative. Model behavior is another critical aspect of AI. While these systems can produce remarkable results, their inner workings often function as a "black box," making it difficult to fully understand how specific decisions are made. This lack of transparency poses challenges, especially in high-stakes applications like healthcare or criminal justice, where accountability and fairness are paramount.

Agentic AI: A Transitional Step Toward General AI?

In the rapidly evolving landscape of artificial intelligence, agentic AI systems have emerged as a central focus and the talk of the day in AI circles as of late 2024 early 2025. This burgeoning field, still in its infancy, represents an exciting yet uncertain frontier. Designed to operate autonomously within defined boundaries, agentic AI systems integrate elements of decision-making, goal-setting, and adaptability. These qualities distinguish them from traditional narrow AI and position them as a critical transitional step toward the possibility of general AI.

What Makes AI "Agentic"?

Agentic AI systems differ from conventional narrow AI in their ability to act as self-directed agents. Rather than simply executing predefined tasks, these systems can:

* Perceive and interpret their environment using sensors, data inputs, or external signals.

* Make autonomous decisions based on dynamic goals or evolving priorities.

* Take deliberate actions to achieve their objectives.

* Learn and adapt from outcomes, refining their performance over time.

A traditional AI in customer service might only respond to questions with pre-written answers. But an agentic AI system could handle conversations on its own, spot patterns in customers' behavior, escalate complex problems, and suggest solutions before they happen. It would also learn from each interaction to get better in the future.

Agentic AI is not just a technological development; it is a moving target, a field that is rapidly changing as researchers and practitioners explore its possibilities and implications. Despite significant progress and incredible enthusiasm, nobody truly knows how agentic AI will unfold. The concepts and applications being developed today could look entirely different a few years from now, as breakthroughs, challenges, and ethical considerations shape its trajectory. This uncertainty makes agentic AI both exhilarating and daunting, a testament to the pace of innovation in artificial intelligence.

Real-World Applications and Their Impacts

Agentic AI is already making its mark across industries. Advanced self-driving cars are one of the most prominent examples, combining real-time perception, decision-making, and navigation to function autonomously. In logistics, agentic AI systems are optimizing supply chain networks by dynamically coordinating between multiple agents to enhance efficiency. Healthcare professionals are exploring these systems to manage complex fluid workflows, ranging from patient monitoring to resource allocation.

These applications highlight the transformative potential of agentic AI while showcasing its limitations and risks. For instance, while agentic systems can operate independently, their decisions may have unintended consequences if their objectives are misaligned, or their adaptability is insufficient.

AI Interaction Skills: Harnessing the Power of Contemporary AI Tools

As AI systems like large language models (LLMs) become more ingrained in our daily lives and work, the development of AI interaction skills has become a crucial competency. These skills act as a bridge between what AI can achieve and how effectively humans can utilize its potential. Whether you're an everyday user relying on user-friendly interfaces (UIs) or a developer delving into APIs and advanced integrations, understanding how to interact with AI tools is vital to maximizing their value.

Principles of Prompt Engineering

At the core of effective AI interaction is prompt engineering, the technique of crafting clear and precise instructions to guide the AI's output. Imagine a prompt as the question or command you give to an AI system; it determines the quality, relevance, and accuracy of the response. For instance, asking a language model, "What are the benefits of renewable energy?" might give you a general answer, while a more specific prompt like "List three environmental benefits of solar power, with supporting data," will produce a more focused and actionable response. In real-world applications, prompt engineering can vary significantly. A marketer might use prompts to generate engaging ad copy, while a software developer could design prompts to debug code. The more structured and clear the prompt, the better the AI performs. Tools like OpenAI's playground allow users to experiment with prompt engineering, demonstrating how minor adjustments can significantly influence results.

Understanding Context and Token Limitations

AI systems like LLMs use tokens to process input and generate output. A token is a piece of text, such as a word or part of a word, and models like GPT have limits on the number of tokens they can handle at once. Understanding these limitations assists users in optimizing interactions.

For example, when summarizing lengthy documents, knowing the token limit ensures the input isn't cut short, which could result in incomplete or misleading outputs.

The context is just as important as token limits. AI models generate responses based on the information they're given within a session, meaning the quality of output heavily depends on how well the context is set. For instance, in customer service, providing an AI chatbot with detailed context about a user's query leads to more accurate and helpful responses. Developers accessing APIs can manage token usage programmatically, allowing for dynamic adjustments for tasks like summarization, analysis, or creative content generation.

Identifying AI Hallucinations and Errors

Despite their capabilities, AI systems are susceptible to hallucinations, confidently producing information that is false or unsupported. For example, an AI might concoct a citation for a scientific paper or generate incorrect details about a historical event. Identifying these errors is a vital skill for ensuring AI outputs are trustworthy. A journalist using an AI tool to draft an article might come across fabricated quotes, while a student could receive false information in an AI-assisted research paper. Identifying when an AI "hallucinates" requires a mix of domain knowledge and critical thinking. For developers, monitoring outputs programmatically and incorporating error-checking algorithms into workflows is essential to maintaining accuracy.

Verification and Fact-Checking Strategies

To counter AI errors, users need to implement robust verification and fact-checking strategies. Cross-referencing AI outputs with credible sources is one method. For example, if an AI suggests that "global carbon emissions decreased by 10% last year," verifying this claim against trusted environmental databases ensures its validity. Real-world applications of fact-checking span industries. A medical professional using AI to review patient symptoms might compare the AI's diagnosis with established medical literature. Similarly, educators incorporating AI-generated content into lesson plans can ensure accuracy by consulting reliable academic resources. Developers often integrate secondary verification layers, such as calling external APIs or databases, to confirm the correctness of

AI-generated outputs.

Cloud Computing and Platform Literacy: Core Requirements for AI Mastery

We first touched on cloud computing in Chapter 4, introducing its fundamental role in modern AI. Here, we will explore it in greater depth, examining how it has transformed the way businesses and individuals interact with technology, particularly its profound impact on artificial intelligence.

While the "cloud" may initially seem like an elusive concept, it is essentially an assortment of distributed computing resources that offer flexible, scalable, and on-demand services. Platforms like AWS (Amazon Web Services), which I've had the opportunity to delve into, are a testament to the intricate and advanced nature of cloud infrastructure. These systems, which have taken billions of dollars and nearly two decades to perfect by a workforce of thousands of engineers, are the backbone of modern AI. Let's explore the fundamentals of cloud computing and its role in an AI-centric world.

Decoding Cloud Computing

At its core, cloud computing provides on-demand access to computing resources via the internet. These resources are categorized into three main service models:

1. Infrastructure as a Service (IaaS): This can be likened to renting the basic elements of computing, such as virtual machines, storage, and networks. AWS Elastic Compute Cloud (EC2) is a prime example, where users can configure and manage their own virtual servers.

2. Platform as a Service (PaaS): PaaS simplifies the development process by offering pre-configured platforms where developers can build, test, and deploy applications. AWS Elastic Beanstalk is an example of this, as it takes care of much of the infrastructure management, allowing developers to concentrate on coding.

3. Software as a Service (SaaS): SaaS delivers fully functional software applications over the internet, eliminating the need for local installation or maintenance. Google Workspace or Salesforce are examples of SaaS platforms that users can access through their browsers.

Understanding these models can help demystify the functioning of cloud services and how they can be utilized to meet both personal and business objectives.

The Power of Distributed Computing

Distributed computing, a characteristic feature of cloud platforms, involves spreading computational tasks across multiple machines. This ensures scalability, fault tolerance, and efficient resource utilization. For instance, when you upload a file to AWS Simple Storage Service (S3), the file is not stored in a single location but is replicated across multiple data centers for reliability. Distributed computing also powers modern AI applications. Training extensive AI models, such as GPT, requires enormous computing resources. Cloud platforms distribute this workload across clusters of GPUs or TPUs (Tensor Processing Units), making the task not only feasible but also efficient.

Exploring Storage and Computing Resources

Cloud platforms offer a wide array of storage and computing resources, tailored to cater to diverse needs. From object storage in AWS S3 for data archiving to high-performance computing on AWS EC2 instances for AI training, the options are virtually limitless. Consider AI development: A data scientist might store a large dataset in AWS S3, use AWS SageMaker to train a machine learning model, and then deploy the model on an EC2 instance for real-time inference. This seamless integration of storage and compute resources makes cloud platforms indispensable for AI workflows.

Understanding Security and Access Management

Security is a critical element in cloud computing, given the sensitive nature of data stored and processed on these platforms. AWS provides robust tools like Identity and Access Management (IAM) to control access to resources. For example, a business can restrict who can access specific databases, ensuring only authorized personnel have the necessary permissions. Beyond access management, cloud platforms also offer encryption for data at rest and in transit, Distributed Denial of Service (DDoS) protection, and compliance with industry standards like the General Data

Protection Regulation (GDPR) and the Health Insurance Portability and Accountability Act (HIPAA). Understanding these security features is crucial for anyone using cloud services, especially in AI, where data privacy and ethical considerations are paramount.

Cost Optimization Principles

One of the most challenging yet rewarding aspects of cloud computing is cost optimization. Cloud platforms operate on a pay-as-you-go model, which means costs can quickly escalate if not managed carefully. Tools like AWS Cost Explorer or AWS Trusted Advisor help users identify unused resources, optimize storage usage, and choose cost-effective configurations. For instance, switching from on-demand EC2 instances to reserved or spot instances can lead to significant cost savings for AI training tasks. Similarly, using lifecycle policies for data in S3 can reduce costs by automatically archiving infrequently accessed files to lower-cost storage tiers like Glacier.

In essence, cloud computing, with its distributed resources and flexible service models, forms the backbone of modern AI and countless other technologies. Platforms like AWS demonstrate the immense power and complexity of the cloud, offering tools and services that simplify even the most demanding tasks. Whether you're training an AI model, storing sensitive data, or deploying an application, understanding the basics of cloud computing unlocks the ability to harness these systems effectively and efficiently. As this field continues to evolve, cloud literacy will remain a critical skill for navigating the AI-driven future.

Mastering Platform Navigation: The Key to Thriving in the AI Ecosystem

The AI landscape is filled with platforms that cater to various needs, from simple applications to complex development environments. OpenAI, for instance, provides user-friendly interfaces for interacting with language models like ChatGPT while offering developer access through APIs for more sophisticated integrations. Google Cloud AI specializes in machine learning with tools like AutoML, enabling users to train custom models without deep technical expertise. Microsoft Azure AI offers capabilities such as natural language processing and computer vision, deeply integrated with other Microsoft services. Each platform has its strengths, and understanding their unique capabilities allows users to select the best tools for their projects. For example, a marketing professional might use Jasper AI for content generation, while a data scientist might prefer TensorFlow on Google Cloud for building custom machine learning models.

Understanding Cloud Service Provider Interfaces

The interfaces developed by cloud service providers like AWS, Google Cloud, and Microsoft Azure are gateways to their extensive capabilities. These platforms typically offer web-based dashboards for accessing services, configuring resources, and monitoring usage. The AWS Management Console, for example, provides a centralized view of services like EC2, S3, and SageMaker, enabling users to launch virtual machines, manage storage, and deploy AI models seamlessly. For those seeking advanced functionality, cloud providers offer command-line interfaces (CLIs) and software development kits (SDKs). These tools enable developers to automate tasks, integrate cloud services into applications, and manage resources programmatically. Learning to navigate both the dashboard and these advanced tools ensures a well-rounded mastery of cloud environments.

Exploring Development Environments and Tools

AI development often requires specialized environments and tools that cater to coding, model training, and deployment. Integrated Development Environments (IDEs) like PyCharm and Visual Studio Code are popular choices, offering features like debugging, syntax highlighting, and plugin support for frameworks like TensorFlow and PyTorch. For machine learning workflows, platforms like Google Colab and Jupyter Notebook provide interactive environments for coding and experimenting with models. These tools make it easy to write and test Python code, visualize data, and document findings in a single interface. Combined with version control and cloud integration, these environments streamline the entire development process.

Understanding Version Control Concepts

In collaborative AI and software projects, version control is essential for managing code changes and ensuring consistency across teams. Platforms like GitHub and GitLab enable users to track modifications, collaborate on codebases remotely, and maintain a history of updates. For example, a team developing an AI application can use Git to manage different branches of the project, such as testing new features or fixing bugs, without disrupting the main codebase. When changes are complete, they can merge these branches back into the primary version, ensuring the project remains cohesive. Version control also protects against data loss and makes it easier to roll back to previous versions if errors are introduced, making it an invaluable tool in any development workflow.

The Power of Integration and Workflow Automation

One of the most transformative aspects of platform navigation is the ability to integrate services and automate workflows. Systems like Zapier, Integromat, and AWS Lambda facilitate the connection between diverse systems, enabling the automation of tasks across multiple platforms. For example, a customer service chatbot powered by AI can integrate with a CRM system to update customer records in real time, ensuring seamless communication across departments. The automation of workflows also applies to model deployment and scaling. Platforms such as Docker and Kubernetes allow developers to place applications into containers, ensuring consistent performance across various environments while automating deployment and resource allocation. This level of integration and automation significantly boosts productivity and reduces manual labor. Navigating platforms involves more than just interacting with tools; it's about mastering an ecosystem that enables you to efficiently and effectively achieve your goals. From understanding the strengths of common AI platforms to leveraging version control and workflow automation, these skills are indispensable to modern technological competence. As AI platforms and cloud services continue to evolve, maintaining proficiency in navigating these systems will ensure you stay adaptable and capable in an increasingly complex digital landscape.

Security Awareness: Safeguarding Yourself and Your Data in a Digital World

In a world increasingly powered by AI and interconnected systems, comprehending the principles of security awareness is not just an optional extra; it's fundamental. Whether managing personal accounts or interacting with complex platforms, the ability to protect your information and digital identity is crucial. This section delves into key concepts in cybersecurity, from defending against social engineering to ensuring data privacy and recovery.

Basic Cybersecurity Principles

Cybersecurity fundamentally revolves around three primary objectives: confidentiality, integrity, and availability (CIA).

• Confidentiality ensures that sensitive information is accessible only to authorized individuals or systems.

• Integrity guarantees that data remains accurate and unaltered, protecting it from tampering or corruption.

• Availability ensures that systems and data are accessible when needed, even in the face of cyberattacks or technical failures.

By understanding and prioritizing these principles, individuals and organizations can build robust defenses against common cyber threats.

Password Management and Authentication

Passwords are a primary defense mechanism, but weak or reused passwords often represent the weakest link in cybersecurity. To strengthen this layer of protection:

• Use password managers like LastPass or Bitwarden to generate and store complex, unique passwords for each account.

• Enable multi-factor authentication (MFA) wherever possible, requiring a second verification step, such as a code sent to your phone or biometric identification.

A robust password policy can significantly reduce the risk of unauthorized access, whether for personal accounts or enterprise-level systems.

Social Engineering Awareness

One of the most insidious threats in cybersecurity is social engineering, where attackers manipulate individuals into divulging confidential information or performing actions that compromise security. Phishing emails, deceptive phone calls, and fake websites are common tactics. For instance, you might receive an email that appears to be from your bank, asking you to verify your account details. A closer look often reveals subtle red flags, like misspelled words or suspicious email addresses. The key to countering these attacks is vigilance, double-checking sources, avoiding unsolicited links, and reporting suspicious activity.

Data Backup and Recovery Concepts

In an era where data is invaluable, losing access to critical files can be devastating, whether due to a cyberattack, hardware failure, or accidental deletion. Data backup and recovery strategies ensure that your information remains safe and retrievable. Some of the common approaches to data backup include:

• Implement regular backups to external drives or cloud services like Google Drive, Dropbox, or AWS S3.

• Use incremental backups to save only the changes made since the last backup, minimizing storage requirements.

• Test your recovery process periodically to ensure you can restore data when needed.

For businesses, automated backup systems integrated with disaster recovery plans are essential to maintain continuity during unexpected disruptions.

Privacy Protection Measures

With the rise of AI and big data, protecting your privacy has never been more important. Personal data, such as location, browsing habits, and communication history, is often collected, stored, and analyzed by various platforms. To safeguard your privacy:

• Use browser extensions like Privacy Badger or uBlock Origin to block trackers and intrusive ads.

• Limit the permissions granted to apps and services, especially those requesting access to sensitive data.

• Regularly review privacy settings on social media platforms and other online accounts.

In addition, understanding how companies use your data, such as reading privacy policies and opting out of unnecessary data collection, empowers you to take control of your digital footprint.

Security awareness is more than just a technical skill; it's a mindset. From managing passwords effectively to recognizing social engineering tactics, and from backing up critical data to protecting personal privacy, these measures form the foundation of digital self-defense. In a world where the lines between the physical and digital blur, staying vigilant and proactive ensures that you remain in control of your information and identity, no matter what challenges arise.

Embracing the Journey of Digital Literacy

This journey through "Digital Literacy and Technical Skills for the AI Era" has been nothing short of massive. The sheer scale of what we've explored, from foundational technical concepts to the intricacies of cloud computing, AI interaction, platform navigation, and security awareness, can feel incomprehensible, daunting, or even outright scary. And that's perfectly okay. It's huge, truly huge. The truth is, nobody fully comprehends it all.

When it comes to digital literacy and understanding of AI, the most important takeaway is this: it's not about mastering everything but about being on the journey. In the fast-evolving digital landscape, maintaining curiosity, embracing ongoing learning, and staying adaptable are far more valuable than memorizing every technical detail. The world of digital literacy and AI isn't a destination; it's a constantly shifting terrain. As long as you're open to exploring new tools, understanding emerging concepts, and adapting to changes, you're already way ahead of the curve. Remember, the goal isn't perfection; it's progress.

By building on what we've explored in this chapter and the previous one, you are equipping yourself with the mindset and skills needed to thrive in an AI-driven world. So, take a deep breath, embrace the uncertainty, and keep moving forward. The journey of digital literacy is not about knowing it all but about "learning enough" to navigate the world with confidence, creativity, and a sense of possibility. As we transition from the foundational understanding of digital literacy to the practical applications of AI, the focus shifts to how we can leverage these incredible tools in our everyday lives. The concepts and skills explored so far have provided a solid base for navigating the digital and AI-driven landscape. Now, it's time to explore the tangible ways AI can enhance our personal and professional lives.

In the next chapter, we'll dive into practical strategies and tools, helping you curate your personal AI toolkit and integrate this transformative technology into every aspect of your daily life in meaningful ways. This is where the journey becomes truly hands-on, bridging knowledge with action.

Chapter 11

Harnessing AI Today for Personal Wellbeing

A Holistic Approach

"Happiness is not a matter of intensity but of balance, order, rhythm, and harmony."

Thomas Merton

Bedrock of a Balanced Life

Throughout my years as a personal consultant, I've noticed that clients often approach me with two main concerns: they yearn for more financial resources and stronger relationships. They frequently express sentiments such as, "I know there are other areas I could improve, but right now, my primary focus is on money and finding the right partner." What they often fail to recognize is that these objectives, while seemingly crucial, are actually the result of a more profound, comprehensive process. Wealth and quality relationships are not things one can directly pursue and obtain. Instead, a specific type of personality and their outward expressions naturally attract them. People who are flourishing in every area of their life radiate an aura of self-assuredness, equilibrium, and stability that attracts opportunities, wealth, and meaningful relationships. This is why focusing exclusively on these outcomes often leads to frustration and disappointment; it's akin to attempting to grasp the reflection of the moon in a pond rather than reaching for the moon itself.

The real effort lies in nurturing a comprehensive sense of well-being, which entails enhancing and balancing a range of interconnected aspects of life. These aspects are relationships, career, finance, health and fitness, entertainment, personal growth, living environment, and family and friends. Each of these elements influences the others, and neglecting one area can disrupt the balance and diminish the overall experience. For instance, unresolved relationship issues can negatively affect career performance, or poor health habits can deplete the energy needed for personal growth.

The secret to unlocking true abundance, whether it's in terms of wealth, relationships, or any other aspiration, is not found in relentless pursuit but in fostering balance across these aspects. This process is deeply personal and unique to each individual. It demands introspection, dedication, and the bravery to make gradual changes that collectively transform one's life. When all these aspects of well-being are harmonized, they create a potent synergy. Relationships thrive not because they are pursued, but because you evolve into the kind of person who naturally attracts meaningful connections. Financial success follows not because you focus solely on wealth, but because you cultivate the skills, mindset, and habits that make you a valuable contributor in the world. Comprehensive well-being is not about isolating goals; it's about nurturing the foundation that makes those goals attainable as a natural outcome of your personal evolution.

The Role of AI in Comprehensive Wellbeing: From Mental Bicycles to Ferraris

Steve Jobs famously portrayed computers as "bicycles for the mind," referencing a 1970s article in Scientific American that compared human efficiency with that of various animals. In their natural state, humans were not the most efficient movers in terms of energy expended per distance travelled. Humans ranked somewhere in the middle of the list when walking. However, when equipped with a bicycle, humans became the most efficient movers of all, surpassing even the most adept creatures in nature. This metaphor encapsulates the transformative power of tools to amplify human abilities. Now, consider this: if base computers are bicycles for the mind, enabling us to pedal farther and faster than ever before, then artificial intelligence is a Ferrari. AI accelerates our ability to think, create, and solve problems, propelling us to levels of efficiency, precision, and insight that were previously unthinkable. Just as a bicycle revolutionized human mobility, AI is revolutionizing how we interact with the world and our own potential.

Amplifying Human Potential Across the Eight Aspects of Wellbeing

The ingenuity of AI lies in its ability to integrate into every aspect of our lives, including the eight aspects of comprehensive well-being. Its potential is not about replacing human effort but about enhancing it, offering tools and strategies to balance and improve all areas simultaneously.

1. Relationships: AI-powered communication tools can assist us in navigating complex interpersonal dynamics, enhance emotional intelligence, and even resolve conflicts through sentiment analysis and mediation algorithms.

2. Career: Personalized learning platforms, career development tools, and AI-powered networking applications open new doors for professional growth.

3. Finance: AI assists us in managing budgets, automating savings, and making smarter investment decisions, providing us a clearer path toward financial stability and freedom.

4. Health and Fitness: AI fitness coaches, health monitoring systems, and personalized nutrition apps enable us to optimize our physical well-being like never before.

5. Entertainment: From curating personalized experiences to enabling creative expression through generative art and music tools, AI enriches our leisure time.

6. Personal Growth: AI serves as a mentor, guiding us through mindfulness practices, learning journeys, and goal-setting strategies tailored to our unique aspirations.

7. Living Environment: Smart home systems powered by AI create harmonious living spaces that adapt to our preferences and promote sustainable practices.

8. Family and Friends: AI-powered tools enhance communication, bridge

distances, and provide insights to nurture deeper connections, fostering stronger bonds with the people who matter most.

The Ferrari Effect: Accelerating Balance and Synergy

AI doesn't just amplify our abilities; it allows us to accelerate toward balance and synergy across these aspects. A well-designed AI tool acts like a co-pilot, analyzing our lives comprehensively and identifying areas for improvement. Imagine an AI system that monitors your physical health, emotional well-being, financial goals, and career aspirations, providing actionable insights to harmonize them. Just as the bicycle transformed human mobility and made us the most efficient travelers, AI is redefining human efficiency in navigating life itself. It offers us a Ferrari-like boost, allowing us to think faster, act smarter, and achieve balance more effectively. However, just as driving a Ferrari requires skill and responsibility, leveraging AI for comprehensive wellbeing requires intention and mindfulness. AI's role is not to dictate but to enhance. By thoughtfully utilizing AI, we can enhance our strengths, address our weaknesses, and fully harness its power to accelerate our journey towards a balanced and fulfilling life. The possibilities are endless, and as we explore AI's potential in the eight aspects of well-being, we open the door to a future where technology and humanity work hand in hand to achieve the extraordinary.

AI-Enhanced Relationships: Bridging Gaps and Building Connections

The concept of using AI in relationships often carries a certain stigma. Many view it as artificial, a departure from the authenticity and spontaneity they associate with meaningful connections. But let's pause for a second and consider: Is it truly artificial, or is it simply a tool, a reflection of human creativity? My own story provides some insight. Over a decade ago, my wife and I met online. At that time, we heavily relied on technology, using Photoshop to enhance our photos, translation apps such as Google Translate to overcome our language barrier, and email exchanges to communicate effectively. It was far from seamless, taking nearly two years to overcome barriers and forge a deep connection. Looking back, I realize that with today's AI tools, that journey could have been a lot faster and smoother, yet no less genuine. AI, when used thoughtfully, doesn't replace human effort; it amplifies it. Traditional in-person dating comes with its own set of limitations. Geographic constraints often limit our options to local social circles, reducing diversity in potential partners. Cultural or linguistic differences can create barriers that discourage meaningful connections. And for many, the anxiety of starting conversations or navigating the early stages of a relationship can feel overwhelming, even if compatibility exists. These limitations often leave people feeling stuck, unable to find the depth of connection they desire.

AI can potentially revolutionize this situation. Modern dating platforms like eHarmony and OkCupid use sophisticated algorithms to analyze user preferences, values, and behaviors, matching people on deeper levels than traditional methods allow. Tinder Passport breaks geographical constraints, enabling users to connect with potential partners worldwide. Language barriers, once insurmountable, are now navigable with tools like DeepL Translator, which offers nuanced and accurate translations for conversations. AI can even mitigate the challenge of initiating conversations. Apps like Hinge provide tailored prompts to help users start meaningful dialogues, while tools like ChatGPT or Claude can assist in crafting thoughtful opening messages. These features help users overcome the initial intimidation factor, allowing them to focus on building genuine connections. AI's potential doesn't end with dating; it extends to enhancing existing relationships as well. Artificial intelligence tools such as Replika provide a safe haven for people to delve into their emotions and modes of communication, promoting personal growth that can have a positive influence on relationships. Platforms like Lasting, designed for couples, offer custom-tailored relationship advice, interactive activities, and strategies for communication.

These resources assist partners in overcoming challenges and strengthening their bond. Sentiment analysis tools, a feature of apps like Tone Analyzer, aid in enhancing clarity and empathy in conversations, reducing potential misunderstandings. The decision to incorporate AI into relationships is a personal one. For those who flourish with traditional methods, there's no need to alter their approach. However, for those facing obstacles, whether they are logistical, cultural, or emotional, AI offers a solution.

As I said before, I say it again. AI does not aim to replace the human aspect but rather to enhance it. In a world where connections are increasingly significant, AI serves not as a replacement but as a bridge. It enables us to establish genuine and rewarding relationships, bolstered by technological support.

AI in Careers: Redefining Opportunities Across Paths

Careers come in a variety of forms, each with its own unique advantages and challenges. Some individuals thrive in the structure and stability of corporate life, advancing in established organizations. Others pursue entrepreneurial opportunities, driven by the thrill of creating something new. Business owners deal with the complexities of managing operations, while freelancers and self-employed individuals establish their niche with greater autonomy. Despite these differences, a common thread connects them all: the need to adapt, grow, and navigate challenges in a constantly evolving landscape.

Traditional career paths, despite their merits, have inherent limitations. Corporate roles can often feel constricting, with opportunities limited by hierarchy or organizational structure. Entrepreneurs face high levels of risk and uncertainty, particularly in competitive markets. Freelancers and self-employed professionals may grapple with inconsistent income, limited access to resources, and the constant pressure to stay relevant. Broadly, time constraints, skill gaps, and market dynamics often impede progress, leaving individuals feeling stuck or underutilized. This is where AI comes in as a transformative force, reshaping career possibilities across every path. For corporate professionals, tools like Microsoft Power BI or Tableau provide actionable insights by analyzing vast amounts of data, enabling employees to make informed decisions faster. Zapier, an automation platform, simplifies workflows by connecting different apps and automating repetitive tasks, freeing up more time for strategic and creative work. Platforms like Grammarly Business enhance communication, ensuring emails, reports, and presentations are polished and professional.

Entrepreneurs benefit immensely from AI tools that reduce guesswork and amplify efficiency. For example, MarketMuse assists with content strategy by identifying gaps in online markets, while Hootsuite uses AI to optimize social media campaigns. For managing customer relationships, platforms like HubSpot CRM offer automated lead tracking, email marketing, and sales insights, helping entrepreneurs scale their ventures with precision.

Freelancers and self-employed professionals are not left behind. Upwork's AI-powered job-matching algorithm connects them with clients aligned to their skills, while Notion provides customizable workspaces to organize tasks and projects efficiently.

Creatives can rely on tools like Canva Pro, which integrates AI for quick graphic design, or Writesonic to generate engaging content effortlessly. AI also bridges skill gaps through platforms like Coursera, Udemy, and edX, offering personalized learning paths tailored to individual needs.

For freelancers or professionals looking to pivot careers, LinkedIn Learning provides AI-recommended courses based on current trends and user profiles, ensuring relevance in the ever-evolving job market.

Additionally, AI-driven collaboration tools like Slack or Trello enable teams across time zones to work cohesively, breaking geographic and professional barriers. These platforms don't just simplify communication; they enhance it with AI-assisted project tracking and automation of routine updates. Tools such as Tableau for data visualization, Jasper AI for content generation, and DALL-E for creative brainstorming create new opportunities for innovation and efficiency. Whether climbing the corporate ladder, launching a business, or thriving in the gig economy, these cutting-edge tools redefine the possibilities of career success. In this rapidly evolving career landscape, partnering with AI isn't just an advantage; it's a necessity. Adopting and mastering these technologies, people can overcome traditional career limitations and position themselves to thrive in the future of work.

AI in Finance: Turning the System to Your Advantage

Finance is often perceived as an intimidating, opaque system, and rightly so, as it wasn't necessarily designed to make individuals wealthy and free. At its core, the financial system is a network of institutions, regulations, and tools aimed at managing the flow of money. While it provides opportunities for growth and stability, it often operates with the interests of corporations, banks, and governments in mind.

The system is complex, with barriers like jargon, access, and financial literacy making it seem inaccessible to the average person. However, finance is not an impenetrable fortress. It's a systematic structure, and like any system, it can be learned and leveraged. Once you understand the rules and mechanics, you can begin to turn the system to your benefit, managing your money more effectively, building wealth, and even achieving financial freedom. The key is understanding and leverage, and this is where AI becomes a transformative ally. AI has the potential to democratize access to financial tools and strategies, breaking down barriers that once required specialized knowledge or expensive advisors. For instance, budgeting apps like Mint or YNAB (You Need A Budget) use AI to analyze your spending habits, offering personalized insights into where your money is going and how you can save more effectively. These tools go beyond traditional spreadsheets by automatically categorizing expenses and identifying areas for optimization, making financial management more intuitive and actionable.

When it comes to investing, AI is a game-changer. Robo-advisors like Betterment and Wealthfront leverage sophisticated algorithms to create personalized investment portfolios tailored to your goals, risk tolerance, and timeline. These platforms continuously monitor market conditions and rebalance your portfolio as needed, ensuring that your investments remain aligned with your objectives. They remove much of the guesswork and emotional decision-making that often undermines individual investors, making it easier to stay disciplined and focused on long-term growth. AI also excels in opportunity analysis. Tools like Seeking Alpha and Zerodha Streak provide AI-driven insights into market trends, enabling individuals to spot opportunities that align with their financial strategies. For small business owners or freelancers, platforms like QuickBooks or Xero integrate AI to manage cash flow, automate invoicing, and forecast financial health, allowing you to make informed decisions with confidence.

Perhaps the most transformative aspect of AI in finance is its ability to level the playing field. Previously, only those who could afford professional services could access advanced financial analysis or personalized advice. Today, AI brings these capabilities to anyone with a smartphone, empowering individuals to not only manage their finances but to strategize for long-term wealth creation. The financial system may not be designed with your freedom in mind, but with the help of AI, you can take control. By leveraging these tools, you can learn the rules, turn them to your advantage, and create a future that aligns with your goals. AI doesn't just make finance easier; it makes financial freedom attainable.

Health and Fitness: Leveraging AI for Enhanced Wellness

Health and fitness, as the foundation of life, influence our energy, productivity, and emotional well-being. However, achieving and maintaining good health and fitness can be challenging in the face of modern life's demands. This is where Artificial Intelligence comes into play, revolutionizing our approach to physical well-being by delivering personalized guidance, motivation, and accountability. Artificial Intelligence has immense potential in the field of health and fitness, providing a comprehensive solution that combines the roles of a personal trainer, nutritionist, and health coach, all within your reach. Apps such as MyFitnessPal and Lose It! utilize AI to analyze your dietary habits, monitor your caloric intake, and suggest healthier alternatives.

These tools go beyond mere calorie counting; they learn your preferences over time, customizing recommendations to fit your lifestyle. Whether your goal is weight loss, muscle gain, or healthier eating, AI provides adaptable, actionable insights.

Fitness is another area where AI excels. Virtual fitness coaches like Fitbod and Future design personalized workout plans based on your fitness level, available equipment, and progress. Unlike one-size-fits-all programs, these AI-driven systems evolve as you do, ensuring your workouts remain challenging and effective. Wearable devices such as Fitbit or WHOOP utilize AI to track your physical activity, heart rate, and sleep patterns, offering real-time feedback to optimize your performance and recovery.

Motivation, often the most challenging aspect of maintaining fitness, is where AI's interactive capabilities truly shine. AI-powered platforms like Noom combine psychological insights with personalized coaching to keep you engaged and committed to your fitness journey. They offer more than simple reminders, providing encouragement, insights into behavioral patterns, and strategies to overcome obstacles.

AI also plays a pivotal role in preventative health. By integrating data from wearables, health apps, and even medical records, AI tools can identify early signs of health issues, recommending actionable steps. For instance, Apple Health or Google Fit can detect irregular heart rhythms or elevated stress levels, prompting users to seek medical advice before problems escalate. The beauty of AI in health and fitness lies in its ability to adapt to individual needs, making the process manageable and enjoyable.

Again, it is not about replacing human effort but enhancing it, providing the structure and guidance needed to stay on track. In a world where time is scarce and distractions abound, AI offers a lifeline to prioritize our health. Leveraging these tools, we can build habits, achieve goals, and maintain the physical well-being that allows us to thrive in all life aspects. After all, when health is taken care of, the possibilities for everything else are endless.

Entertainment: Enriching Leisure with AI While Maintaining Control

Entertainment is a broad and diverse domain, encompassing everything from passive activities like watching movies or listening to music to gaming, hobbies, or creative expression. However, in today's world of infinite choices and constant distractions, finding the right kind of entertainment that genuinely enriches our lives can be daunting. Here, AI steps in, not only curating personalized experiences but also helping us maintain balance and control.

AI has revolutionized entertainment in incredible ways. Streaming platforms like Netflix and Spotify leverage machine learning to analyze our preferences, delivering tailored recommendations that align with our tastes. These algorithms sift through vast libraries of content, introducing us to new shows, songs, or genres we might never have discovered on our own.

Furthermore, generative AI tools like DALL-E, Sora, and Midjourney for art or AIVA for composing music enable us to become creators, transforming leisure time into opportunities for creative self-expression. AI-powered video games also adapt dynamically to player skill levels, making gameplay more engaging and immersive.

Yet, the transformative power of AI in entertainment doesn't end with curation or creativity. It's also a powerful tool for enhancing hobbies and active pursuits. For instance, platforms like MasterClass use AI to recommend lessons from world-class experts tailored to your interests, while tools like Skillshare guide you toward learning and refining creative skills.

For those who love photography, AI-driven photo editing apps like Luminar AI simplify complex adjustments, enabling users to create stunning visuals with ease.

However, with great access comes great responsibility. Entertainment, by its very nature, can be all-consuming, and the abundance of options AI provides can make it easy to lose track of time or prioritize passive over meaningful engagement. This is where we must exercise wisdom and use AI not just to enhance entertainment but to set boundaries. Tools like Freedom and StayFocusd allow us to limit access to distracting apps or websites, while features built into devices, like Apple's Screen Time, provide insights into usage patterns and encourage balance.

Quality entertainment is more than just convenience or indulgence; it's about choosing activities that truly enrich our limited leisure time. AI can assist by narrowing options to those that align with our values, goals, and energy levels, ensuring we derive maximum joy and inspiration from every moment. But the key lies in remaining in charge of the experience, using AI as a guide rather than letting it dictate our choices. Entertainment has the power to uplift and rejuvenate, but without mindfulness, it can also become a source of distraction or dependency. Leveraging AI thoughtfully, we can curate high-quality experiences that enrich our lives while maintaining the discipline to stay balanced. In this way, AI becomes not just a tool for leisure but a partner in ensuring that the time we spend enjoying life is both fulfilling and intentional.

Personal Growth: AI as a Guide to Expanding Horizons

Personal growth is a lifelong journey, encompassing everything from acquiring new skills to cultivating mindfulness, from broadening worldviews to exploring spiritual practices. It's a deeply individual process, often driven by curiosity, reflection, and the desire to better understand oneself and our environment.

In today's world, AI has proved to be a powerful ally in this journey, offering tools and strategies that can tailor personal development to our unique aspirations. However, like any tool, AI requires deliberate and purposeful use to ensure it enhances rather than distracts from our personal growth. It's essential to develop the discernment needed to separate genuine value from misleading ideologies, false information, deliberate misinformation, and other harmful influences that can derail progress.

One of the most exciting areas of personal growth is the ability to extend our worldviews. AI-powered platforms like Coursera and Duolingo make learning new subjects and languages accessible to anyone with an internet connection. These tools go beyond traditional methods, using adaptive learning techniques to personalize lessons based on your pace and preferences. For instance, Duolingo gamifies language learning, keeping users motivated while AI ensures lessons are challenging but achievable. Similarly, apps like BBC Maestro introduce you to thought leaders across various disciplines, offering insights that challenge and expand your perspectives. Generative AI tools like Claude Sonnet, Quillbot, Sora and Midjourney and many others provide hands-on opportunities to experiment with writing or art, turning abstract interests into tangible creations. These tools make it easier than ever to explore new facets of yourself and translate curiosity into action.

For those seeking spiritual growth or mindfulness practices, AI offers guidance in creating personalized routines and fostering self-awareness. Apps like Headspace and Calm use AI to tailor meditation and mindfulness practices, offering sessions that align with your emotional state and goals. Tools such as Insight Timer surpass this by connecting users with global communities and utilizing AI to recommend practices that best suit individual needs.

AI technologies can make age-old practices more accessible and compatible with modern living. However, as previously discussed, it's crucial to apply discernment. The data it's given determines AI's accuracy, and not all of its suggestions will meet your goals. Growth is not a competition, and while AI can enrich your journey, it falls on you to contemplate what truly benefits your long-term development. AI doesn't replace the work of growth; it simply makes the process more accessible, engaging, and personalized, empowering us to become the best versions of ourselves.

Living Environment: Enhancing Our Lives with AI

Our surroundings greatly influence who we are. Our neighborhoods, homes, and immediate environments act as a silent guide, subtly shaping our behaviors, thoughts, and feelings. A vibrant, well-maintained neighborhood can inspire creativity and optimism, while a disordered or poorly kept environment may lead to stress and stagnation.

The living environment is not merely a backdrop to our lives; it's an active participant, influencing how we feel, work, and grow. Consider the power of a well-designed room or a home, where every corner is purposeful, and every element mirrors who you are. Such a space doesn't merely shelter you; it nurtures you. It provides the mental clarity and emotional stability that allows you to flourish.

Conversely, limitations in our environment, whether due to lack of resources, space, or external factors, can become barriers that restrict our potential. Here, AI is again offering innovative ways to not only enhance our physical spaces but also create entirely new dimensions of living. AI-powered smart home systems are revolutionizing how we interact with our environment. Devices like Amazon Alexa or Google Home allow seamless control over lighting, temperature, and entertainment, adapting to our preferences and routines. Nest Thermostats learn your habits to optimize energy use, striking a balance between comfort and sustainability.

These technologies not only make life more convenient; they create environments that actively promote our well-being. For those living in constrained spaces or with limited resources, AI offers solutions that can make even small areas more functional and inviting. Apps like Houzz or Room Planner use AI to help you visualize and design your living space, maximizing utility without compromising aesthetics.

But AI doesn't stop at physical spaces; it also extends into virtual environments. Tools like Spatial or Horizon Workrooms, AI-powered virtual reality platforms, allow us to create digital spaces for collaboration, relaxation, or creative exploration. These virtual environments can serve as an extension of our physical homes, offering new opportunities for interaction and self-expression, particularly for those working or socializing remotely.

AI also plays a role in promoting sustainability within our living environments. Smart energy management systems, such as Tesla Powerwall or EcoBee, help monitor and reduce energy consumption. Even urban planning tools powered by AI are shaping greener, more efficient neighborhoods, ensuring that the collective environment benefits individuals and communities alike.

Ultimately, our living environment reflects who we are and is a foundation for who we can become. AI has the potential to enhance this relationship, turning our spaces into active participants in our growth and well-being. The key, as always, is to use these tools thoughtfully, creating spaces that not only adapt to our needs but elevate the lives we aspire to live.

Family and Friends: The Pillars of Our Identity

If our environment profoundly shapes us, our family and friends shape us even more. From the moment we are born, our identity begins to form in response to how those around us perceive and interact with us. A child's first understanding of themselves is mirrored back through the reactions of their family, whether their expressions are met with support or rejection, encouragement or criticism. These early interactions lay the foundation for self-worth, confidence, and emotional resilience. While this influence is most defining in childhood, it doesn't stop when we reach adulthood. The opinions, behaviors, and emotional availability of family and friends continue to affect us throughout our lives. Even as we develop independence, their actions and words can either nurture our growth or become obstacles we must navigate. These relationships remain central to our sense of belonging and purpose, shaping how we perceive the world and ourselves within it.

When it comes to family, we are given these relationships, not chosen. While we can't pick our relatives, we can choose how we interact with them. Improving family dynamics often involves fostering communication, understanding, and empathy. AI-powered tools like Lasting offer guided exercises and insights to help families resolve conflicts and strengthen bonds. These tools can suggest ways to express gratitude or actively listen during difficult conversations, creating opportunities for healing and connection.

Friends, on the other hand, are relationships we select, and in today's interconnected world, the concept of friendship has expanded to include virtual connections. AI can play a significant role in helping us cultivate and maintain friendships. Platforms like Meetup or Bumble BFF use algorithms to match people with shared interests, fostering meaningful connections beyond geographical or social constraints. Virtual friends, while not a replacement for human interaction, can provide companionship and understanding when traditional connections feel out of reach.

Apps like Replika create AI-powered companions that engage in supportive and empathetic conversations, offering a sense of connection during challenging times. AI also assists in sustaining long-distance friendships and family relationships. Tools like Marco Polo Video Messenger and Telegram use AI features to optimize communication, enabling seamless video and voice messaging that feels personal despite physical distance.

Language barriers, which once hindered global connections, are now bridged by real-time AI translation tools, allowing families and friends from different cultures to connect effortlessly. While AI provides valuable tools to enhance relationships, it's crucial to approach these technologies with intention. Family and friends remain the most profound influences in our lives, shaping our growth, happiness, and resilience. Using AI thoughtfully, we can strengthen these bonds, create new ones, and ensure that our relationships, both given and chosen, continue to nurture and define us in meaningful ways.

A Personalized Journey

As we conclude this discussion, it's important to reflect on the role AI plays in our personal journeys. Leveraging AI for well-being, as I emphasize again and again, is not about replacing human effort; it's about enhancing it. AI is a tool, a collaborator, and a partner, designed to support us in ways that align with our unique goals and circumstances. But as advanced and transformative as AI can be, it will never replace the deeply human elements that define our lives. AI cannot fill the void of an intimate soulmate or build the bonds that come from shared vulnerability and trust with another person. It can suggest matches, analyze compatibility, or even help improve communication, but the depth of human connection is something only we can cultivate. Similarly, AI will not deliver the satisfaction of an amazing career. It can streamline tasks, optimize workflows, or suggest paths to success, but the passion, dedication, and sense of purpose that come from meaningful work must come from within.

Again, AI cannot deposit a million dollars into our bank accounts or pay our bills. It can assist with budgeting, investment strategies, and financial planning, but financial freedom requires intentionality, discipline, and effort on our part.

Artificial Intelligence is not capable of physically exercising for us, managing our internal bodily functions, or assuming the responsibility of our physical well-being. It can, nevertheless, provide us with customized fitness plans, monitor our progress, and even remind us to remain active. However, the onus to act on these recommendations is on us.

AI can provide us with endless entertainment, curating a continuous stream of content based on our preferences. Yet again, it cannot instill the self-discipline required to disengage when it becomes a potential distraction.

Personal development also remains a deeply individual endeavor. AI can supply resources, insights, and guidance, but real growth stems from introspection, action, and tenacity. It cannot construct a home for us, surround us with a loving family, or create a network of influential friends. These outcomes still depend on our choices, actions, and the relationships we cultivate.

All eight aspects of well-being, relationships, career, finance, health, entertainment, personal growth, living environment, and family are interconnected. To truly flourish, we must nurture them concurrently, balancing the tools AI provides with our own effort, intuition, and control.

As you progress, remember that AI is here to enhance what is uniquely human: our creative passion, empathy, and resilience. Use it to alleviate your burdens, discover new opportunities, and ignite ideas, but let your own aspirations and values remain at the heart of your journey. Together, you and AI can create a life that is not only balanced and fulfilling but also deeply and authentically your own.

As we conclude our exploration of how AI can enhance personal well-being across the eight dimensions of life, it's time to shift our focus outward. The same transformative power that AI offers to individuals can also fuel entrepreneurial ventures, unlocking opportunities once thought unattainable. In the next chapter, we'll delve into how AI is reshaping the business landscape, creating new possibilities for those willing to harness its potential. From niche services that complement AI, such as consulting and training, to entirely new business models driven by personalized content and data insights, AI empowers entrepreneurs to innovate and compete at unprecedented levels. Whether you're a seasoned business owner, a solopreneur, or someone with an idea waiting to be realized, the entrepreneurial opportunities in the AI landscape are vast and waiting to be explored.

Chapter 12

Entrepreneurial Opportunities in the AI Landscape

Exploring New Frontiers

"The secret of change is to focus all of your energy not on fighting the old, but on building the new."

Socrates

A New Era of Entrepreneurship

As we transition from the realm of personal development and wellbeing, let's shift our attention to a field that has been radically altered by artificial intelligence: entrepreneurship. Once a sphere reserved for those with substantial capital, specific expertise, or established networks, the realm of business creation and innovation is now accessible to a much broader audience. AI has diminished many of the traditional obstacles to entry, providing tools that allow individuals to turn ideas into tangible outcomes, automate intricate procedures, and compete with larger corporations in an unprecedented way. However, this newfound accessibility comes with its own set of challenges.

With the lowering of entry barriers, the entrepreneurial landscape has become more saturated and competitive. The capability to start a business is no longer the only differentiator; success now hinges on creativity, adaptability, and a strategic approach to exploiting AI's potential. So, while AI has democratized access to entrepreneurship, it hasn't made the journey effortless. Our aim in this chapter is not to suggest that entrepreneurship has become a simple pursuit, but rather to highlight that it is now feasible for many to enter a field that was previously out of reach.

Whether you're an aspiring entrepreneur with limited resources or an established business owner seeking to innovate, AI presents unparalleled opportunities to scale ideas, develop niche services, and create business models tailored to today's dynamic markets. In the forthcoming pages, we'll delve into how AI can empower you to surmount traditional barriers, unlock innovative business ideas, and flourish in this competitive yet thrilling landscape. AI doesn't guarantee success, but it does level the playing field, providing those with vision and determination the tools to turn potential into reality.

What Is Entrepreneurship?

People often misunderstand entrepreneurship as just running a business. However, it is far more nuanced. At its heart, entrepreneurship is about identifying opportunities, taking calculated risks, and creating something new, be it a product, service, or an entirely novel way of doing things. It necessitates an innovative mindset and a readiness to step into the unknown, often with the aim of solving problems or addressing unmet market needs. While every entrepreneur is a businessperson, not every businessperson is an entrepreneur. The key distinction lies in their approach and intent. A businessperson generally operates within established systems, focusing on optimizing existing processes, scaling a proven model, or delivering value through efficiency. Their goals are often tied to stability and incremental growth. Entrepreneurs, conversely, thrive on disruption. They seek to challenge norms, create new markets, and take risks that often carry the potential for high rewards or significant failure. They embrace the responsibility, and the uncertainty, of forging their own path. The traits that define an entrepreneur often include creativity, resilience, and an appetite for risk.

Entrepreneurs are fundamentally problem-solvers who perceive obstacles not as barriers but as opportunities for innovation. They are comfortable with ambiguity and possess the drive to learn and adapt as they go. Perhaps most importantly, they are propelled by a sense of purpose, a desire to create an impact, whether that's through building a sustainable business, introducing a game-changing product, or addressing societal challenges. Entrepreneurship is not confined to any one industry or demographic. It can be a solo freelancer launching a niche service, a team building a tech startup, or even someone reimagining an existing business to meet new demands.

In today's world, fueled by AI and other technological advancements, the definition of entrepreneurship is expanding, offering more avenues for innovation than ever before. Understanding these distinctions helps clarify the unique mindset and qualities required to embrace entrepreneurship. It's about more than financial gain; it's about creating, innovating, and daring to do what hasn't been done before. As we explore the opportunities AI brings to the entrepreneurial landscape, these foundational ideas will guide us in understanding how this transformative technology is shaping the future of business and entrepreneurship.

Key Democratizing Factors in AI-Powered Entrepreneurship

AI's rise has not only broadened the scope of what entrepreneurs can achieve but has also fundamentally altered who can participate in the entrepreneurial space. Once restricted to those with extensive technical expertise or access to significant resources, AI has become increasingly accessible, thanks to a handful of key factors that have democratized its use. These factors are enabling more people than ever before to bring their ideas to life and compete in markets that were once out of reach. One of the most transformative developments is the availability of cloud-based AI services. Platforms like AWS, Google Cloud, and Microsoft Azure provide powerful AI capabilities that are hosted and maintained remotely, meaning entrepreneurs no longer need to invest in expensive infrastructure or manage complex setups. These services offer everything from natural language processing to image recognition, all accessible through intuitive online dashboards. Entrepreneurs can start using cutting-edge AI tools with just a few clicks, without needing a background in IT or computer science.

Another critical enabler is the shift to pay-as-you-go pricing models. In the past, accessing advanced technology required significant upfront investments, creating barriers for those without deep financial resources. Today, AI services charge users based on what they use, making it possible for small businesses and solo entrepreneurs to experiment and grow without committing to massive costs. This model encourages innovation, allowing users to scale their expenses alongside their successes.

The evolution of user-friendly interfaces has further expanded access. Tools that once required complex coding and specialized knowledge are now designed to be intuitive and approachable. Platforms like Hugging Face and no-code tools like Bubble and Zapier empower entrepreneurs to build AI-integrated applications, workflows, and automations that were previously the domain of skilled developers.

Equally transformative is the way physical infrastructure needs have changed. In the past, starting a new venture often required leasing office space, signing long-term contracts, and committing significant resources to physical locations. These commitments posed enormous financial risks, especially for new entrepreneurs. Today, the virtual office has replaced the traditional one. With cloud-based platforms, shared collaboration tools like Slack and Microsoft Teams, and project management systems like Asana and Trello, entrepreneurs can operate entirely online. Whether working from a garage, a coworking space, or even a café, the need for expensive office infrastructure has been virtually eliminated.

The open-source movement and pre-trained AI models have further lowered barriers. Open frameworks like TensorFlow and pre-trained systems like OpenAI's GPT or Google's BERT allow entrepreneurs to implement sophisticated AI capabilities without needing to build from scratch. These tools drastically reduce development time and costs, enabling businesses to focus on delivering value rather than reinventing the wheel. Taken together, these democratizing factors; cloud-based AI, flexible pricing, user-friendly tools, virtual offices, and pre-built models, have transformed the entrepreneurial landscape. Starting a new venture no longer requires substantial financial commitments, technical expertise, or access to exclusive resources. Instead, anyone with a vision and determination can leverage AI to launch innovative ventures, making entrepreneurship more accessible and inclusive than ever before. As I said before, this shift doesn't make entrepreneurship effortless, but it does make it possible for far more people to participate, dream big, and compete on a global scale.

Real-World Lessons from AI-Driven Success

The entrepreneurial landscape today is a dynamic interplay between innovation, vision, and technology. As we delve into the following case studies, our focus is twofold: first, to understand how AI can empower startups to operate efficiently and effectively, and second, to explore how AI can be seamlessly integrated into the very core of a product or service. These real-world examples offer invaluable lessons for aspiring entrepreneurs navigating the AI landscape. They demonstrate that success often comes from identifying an unmet need, leveraging AI's capabilities to address it, and creating solutions that resonate with users. Whether AI is used as a behind-the-scenes enabler, streamlining workflows, automating processes, and scaling operations, or as the heart of a business's offering, driving value directly to the customer, its transformative potential is undeniable.

These examples demonstrate AI's dual role as both a tool and a product, illustrating its versatility in fostering innovation. Our goal in examining these instances is to extract the principles that contributed to their success. This provides practical insights for those seeking to utilize AI in their entrepreneurial endeavors. These aren't mere abstract theories or hypothetical situations; they are rooted in the practical realities of initiating and growing businesses in today's rapidly changing marketplace. Through these case studies, we witness how AI has expanded the realm of possibilities, opening avenues for creativity, efficiency, and effective problem-solving. Let's delve into how these pioneers have utilized AI to achieve extraordinary success and what lessons we can derive for our own paths.

Case Study 1: Jake Heller and Casetext

Jake Heller's entrepreneurial journey with Casetext is a classic example of blending domain expertise, foresight, and technological innovation to meet a specific market demand. Established in 2013, Casetext stemmed from Heller's firsthand experience as a litigator. He pinpointed a significant gap in the legal research process, a sector ripe for modernization through technology. Initially, Casetext concentrated on developing a platform for legal professionals to access and share legal documents. However, as AI technology progressed, Heller recognized the potential to incorporate advanced AI tools to further revolutionize the industry. This led to the creation of CoCounsel, one of the first AI-powered legal assistants built on OpenAI's GPT-4. CoCounsel wasn't merely an impressive innovation; it was a targeted solution for specific challenges in legal research, providing insights, document summaries, and case recommendations with unparalleled efficiency. This sharp focus on resolving real-world issues solidified Casetext's position as a leader in the legal tech space. By 2023, the company's strategic shift toward AI-powered tools paid off when Thomson Reuters acquired Casetext for $650 million in cash.

Key takeaways from Heller's journey include:

1. Timing Is Critical: Identifying the opportunity to modernize legal research before AI's mainstream adoption allowed Casetext to establish itself early.

2. Domain Expertise Matters: Heller's experience as a litigator gave him profound insights into the needs and challenges of legal professionals.

3. Focus on Real Problems: Casetext succeeded because it applied AI to address genuine inefficiencies instead of pursuing technology for its own sake.

4. Strategic Adaptation: The shift from a document-sharing platform to AI-enhanced solutions demonstrated agility and a sharp understanding of market dynamics.

Case Study 2: Whitney Wolfe Herd and Bumble

Whitney Wolfe Herd, the founder of Bumble, disrupted the saturated online dating industry with a unique twist: empowering women to make the first move. Introduced in 2014, Bumble's innovative approach distinguished it in a crowded market, prioritizing safety, respect, and equality in online dating. Bumble's application of AI has been a significant factor in its success. The platform uses AI to detect and filter out inappropriate behavior, flagging offensive messages to maintain a safe environment. Additionally, Bumble utilizes machine learning to personalize recommendations based on user behavior, ensuring better matches and a more engaging user experience. By focusing on a specific niche and leveraging AI to enhance safety and satisfaction, Bumble experienced rapid growth, amassing millions of users worldwide. Whitney Wolfe Herd became the youngest female CEO to take a company public when Bumble debuted on the NASDAQ in 2021 with a valuation of over $8 billion.

Key lessons include:

1. Niche Focus: Identifying an underserved demographic and tailoring the platform to meet their needs was a game-changer.

2. AI for Safety and Personalization: Leveraging AI to improve user experience and create a safe environment resonated with users and built trust.

3. Brand Differentiation: Bumble succeeded by standing out in a crowded market, offering a fresh approach to online dating.

Case Study 3: Melanie Perkins and Canva

Melanie Perkins co-founded Canva in 2012 with a vision to democratize design. She realized that traditional design tools like Photoshop were too complex and expensive for many users. Her goal was to create a platform that allowed anyone, regardless of skill level, to create professional-quality designs. Canva's success lies in its user-friendly interface, AI-driven features, and accessibility. The platform's AI tools, like automated layout suggestions, background removal, and instant resizing, empower users to create stunning visuals with minimal effort. Canva has also leveraged machine learning to improve its recommendation engine, helping users find templates, graphics, and fonts that suit their needs. Today, Canva is valued at over $26 billion, with more than 100 million active users globally. Its intuitive design and AI enhancements have made it a go-to tool for small businesses, educators, and individuals alike.

Key lessons include:

1. Democratizing Access: Simplifying design tools opened up a market previously dominated by professionals.

2. User-Centric Design: Prioritizing ease of use and leveraging AI to streamline workflows made Canva appealing to a broad audience.

3. Scalability: Canva's model demonstrated how accessible tools can capture massive markets across industries.

Lessons from the Case Studies: Reducing Risk and Amplifying Potential

The success stories of Casetext, Bumble, and Canva offer a wealth of insights into modern entrepreneurship, but one of the most significant lessons is the role AI plays in reducing risk and accelerating innovation. Concepts like Proof of Concept (POC) and Minimum Viable Product (MVP) have long been staples of entrepreneurial strategy, but AI has redefined how these concepts can be applied, making them faster, cheaper, and more accessible. Traditionally, developing a POC or MVP required significant investment. Testing an idea often meant building physical prototypes, setting up factories, or creating full-scale business models, commitments that carried enormous financial and logistical risks. Today, AI and virtual environments have revolutionized this process. Entrepreneurs can now test their ideas digitally, simulating real-world conditions without the need for costly infrastructure. For example, AI tools like SimScale or Autodesk Fusion 360 allow businesses to run virtual simulations of products, testing everything from durability to efficiency before a single physical component is manufactured. Similarly, the concept of an MVP has evolved with AI's capabilities. Platforms like No-Code AI tools (such as Bubble or Zapier) enable entrepreneurs to create functional prototypes of apps, services, or workflows without extensive programming knowledge. These prototypes can be tested with real users, providing valuable feedback and insights with minimal upfront investment. AI-powered analytics tools then help refine these MVPs by analyzing user behavior, identifying pain points, and suggesting improvements.

AI also brings additional leverage to the traditional entrepreneurial toolkit. For instance:

• Market Research Acceleration: Tools like SurveyMonkey AI or Google Trends enable entrepreneurs to gauge market demand, customer preferences, and competitor positioning at an unprecedented scale and speed.

• A/B Testing at Scale: AI platforms like Optimizely allow for simultaneous testing of multiple variables, ensuring entrepreneurs can fine-tune their offerings with precision.

• Iterative Learning Loops: Machine learning models, such as those provided by AWS SageMaker, allow businesses to refine products and services continuously, based on real-time user data.

The ability to leverage these tools has significantly lowered the barriers to entry for entrepreneurs. Ideas that once seemed too risky or resource-intensive to pursue can now be tested and refined virtually, with AI shouldering much of the burden. This has unlocked a wave of innovation, enabling countless individuals to bring their ideas to life. By reducing risk and increasing adaptability, AI doesn't just make entrepreneurship more accessible; it creates a fertile ground for bold experimentation. Entrepreneurs can afford to think bigger and act faster, confident that their ideas can be validated and iterated upon in a controlled, data-driven manner. The result is a new era of entrepreneurship, where creativity and technology converge to turn visions into reality with unprecedented efficiency.

The success stories we've explored, billion-dollar startups like Casetext, Bumble, and Canva, may seem like exceptions to the rule, outliers that only a select few can replicate. And to an extent, that's true. Not everyone will launch a company that reshapes an industry or attracts multi-million-dollar investments. But the principles and opportunities behind these stories are not exclusive to those at the top of the entrepreneurial ladder. One way or another, everyone has the potential to improve their current situation using AI, regardless of their starting point.

For example, certain enterprises are marketing educational resources that promise to furnish individuals with the expertise to operate AI consultancies within a mere three months. These consultancies would assist companies in incorporating AI into their operations. Personally, I view such assertions of accelerated expertise with a degree of skepticism. However, it is undeniable that a select few exceptional individuals might indeed provide tangible value to others through a compressed learning trajectory. For my part, I have opted for a more gradual approach. Over the previous three years, I have devoted an average of 8–10 hours daily, and up to 12–14 hours during periods of examination, to comprehensively understand AI. This commitment is not merely professional; it stems from a genuine curiosity and an aspiration to comprehend the philosophical and technical subtleties of this revolutionary technology. As an old school businessperson, I am a proponent of providing high-value solutions and fostering long-term success. Presently, in addition to conducting research on industry-specific AI applications, I operate an AI consultancy that designs and implements AI systems to automate workflows across various industries. This journey has imparted a crucial insight: AI is a vast field, and it is impossible for anyone to master it entirely. It is tempting to perceive AI as a tool that enables you to do everything. However, the truth is that while AI empowers us to address an almost limitless array of problems, no individual or enterprise can solve them all. Achieving success with AI necessitates concentration on a single problem or domain, and tailoring solutions to meet specific needs. This focus requires not only technical proficiency but also collaboration.

AI flourishes in an environment of partnerships. A technical professional, such as a machine learning engineer, can produce remarkable outcomes when they collaborate with a domain expert who possesses a deep understanding of a specific industry. Similarly, a non-technical entrepreneur can make a significant impact by partnering with a skilled AI solution architect. Cooperation, rather than isolation, is the key to unleashing the full potential of AI. This collaborative mindset should extend beyond professional partnerships. It involves thinking in terms of generating value, not only for clients but also for your collaborators. Interpersonal skills are as important as technical capabilities. Sharing knowledge, building trust, and promoting teamwork are essential to the successful application of AI in the real world. Therefore, while the grand success stories may appear unreachable, they serve as inspiration for what is possible. AI provides tools and opportunities that anyone can utilize to enhance their career, business, or entrepreneurial journey. Regardless of whether you're starting small or aiming high, the journey begins with a focus on addressing meaningful problems and a readiness to collaborate.

Specialized Services in Conjunction with AI

AI Implementation Consulting

For readers who have absorbed the mindset and learning journey we have explored in this book, AI presents a multitude of opportunities for creating value. One particularly promising pathway is providing specialized services that work in tandem with AI, such as AI implementation consulting. This realm is not about constructing AI systems from the ground up but rather about assisting organizations in navigating the intricate and ever-changing terrain of AI technologies to unlock their potential. AI implementation consulting involves directing organizations in their selection, integration, and effective use of AI tools. Given the vast array of AI platforms, applications, and solutions on offer, many businesses find themselves overwhelmed by the choices. This is where consultants come in, bridging the divide between technical possibilities and practical business needs. For instance, a restaurant chain might seek advice on using AI to forecast customer demand or streamline supply chain logistics. A proficient consultant can evaluate their needs and suggest tools like predictive analytics platforms or AI-driven scheduling systems, ensuring the technology aligns with their objectives. Integration is another crucial area of focus.

Many businesses already have established workflows and operational systems, and introducing AI requires meticulous planning to prevent disruption. For example, integrating a chatbot into customer service operations is not simply about installing software; it involves mapping the bot's functions to the existing team's roles, ensuring seamless collaboration between humans and machines. Consultants provide the technical expertise and strategic insight to make this process smooth, minimizing downtime and maximizing return on investment.

Beyond technical aspects, AI implementation consulting also encompasses strategy development and road mapping. Businesses often know they need AI but struggle to chart a clear course forward. Consultants assist them in setting realistic goals, prioritizing initiatives, and crafting actionable plans. For instance, a retail company might start with basic AI applications like personalized recommendations and gradually scale to more advanced systems like inventory optimization. A well-designed roadmap ensures resources are allocated wisely and progress is trackable.

Change management and employee training are equally vital. One of the most significant barriers to successful AI adoption is resistance from employees who fear automation will replace their roles. Consultants play a crucial role in addressing these concerns, emphasizing AI as a tool for augmentation rather than replacement. By providing tailored training programs, consultants enable teams to work effectively alongside AI, ensuring a smooth transition and fostering a culture of innovation.

The value of AI implementation consulting lies in its versatility. It is not limited to large enterprises; small and medium-sized businesses are increasingly looking to adopt AI but often lack the expertise to do so. Consultants who can demystify AI and tailor solutions to specific industries will find themselves in high demand. This field is not about being the most technical person in the room; it's about understanding both the technology and the human factors involved. If you've committed yourself to learning AI concepts and cultivating collaborative skills, AI implementation consulting could be a highly rewarding way to create value, helping businesses harness the power of AI to achieve their goals.

AI Output Verification and Enhancement

As AI continues to generate a staggering amount of content across industries, ranging from automated reports and chatbot interactions to creative output like videos and articles, the need for AI output verification and enhancement has grown exponentially. While AI tools are powerful and efficient, they are not infallible. Errors, biases, and inaccuracies can slip through, making quality assurance and refinement vital services that add immense value to businesses relying on AI-generated content. One of the most significant roles in this space is quality assurance for AI-generated content. AI tools like ChatGPT or DALL·E can generate text, images, and even code, but the outputs are only as good as the data they were trained on and the prompts they receive. This creates a gap where AI's results need to be evaluated for relevance, coherence, and appropriateness. For example, a marketing agency using AI to generate product descriptions might need someone to ensure the tone aligns with the brand's voice and that no inaccuracies undermine credibility.

Fact-checking and accuracy verification is another crucial service. AI systems often produce outputs that appear plausible but are factually incorrect, a phenomenon commonly referred to as "AI hallucination." In fields like journalism, education, or law, where accuracy is paramount, the role of a verifier becomes indispensable. Specialists in this area can cross-reference AI-generated outputs with reliable sources, ensuring that errors are caught before they reach the public. In addition to verification, content refinement and customization add value by tailoring AI outputs to specific contexts. AI can produce a rough draft of an article, a legal brief, or a business proposal, but refining it to meet nuanced requirements often requires a human touch. Consider the company that uses AI to generate a business pitch deck but rely on a consultant to enhance the messaging, adapt it for the target audience, and ensure cultural sensitivity when pitching to international stakeholders.

Compliance and risk assessment services are particularly crucial in regulated industries like healthcare, finance, and legal services. AI-generated content must often meet strict compliance standards, such as adhering to HIPAA in healthcare or GDPR in data privacy. Professionals who understand these regulations can review AI outputs to ensure they meet the necessary legal and ethical requirements. This service not only mitigates risk but also builds trust with clients and stakeholders. The opportunities in AI output verification and enhancement are vast because this role sits at the intersection of technology and human judgment. Despite AI's unparalleled efficiency, it still lacks the nuanced understanding and critical thinking that humans contribute. By bridging this gap, professionals can help businesses maintain the reliability, accuracy, and customization required to succeed in competitive markets. For individuals looking to carve out a niche in this area, a blend of analytical skills, industry-specific knowledge, and a willingness to collaborate with AI tools is essential. This isn't about replacing AI; it's about partnering with it to produce outcomes that are polished, reliable, and impactful. As the use of AI in content generation continues to grow, so too will the demand for those who can ensure its outputs meet the highest standards.

Personalized Content Creation

AI has revolutionized content creation in a world where personalization is crucial for capturing attention and engagement. The ability to tailor content to specific audiences at scale was once the domain of large corporations with dedicated teams of marketers, designers, and writers. Today, AI enables businesses of all sizes, and even individual entrepreneurs, to produce personalized, high-quality content with unprecedented efficiency. AI-assisted writing services are one of the most popular and accessible applications in this field. Tools like ChatGPT and Jasper AI empower creators to draft articles, blog posts, product descriptions, and even technical documentation in minutes. Beyond generating initial drafts, these tools allow users to refine tone, style, and format to suit their target audience.

Beyond text, AI is revolutionizing custom video and image generation. Platforms like Sora, DALL·E, Runway ML, and Pictory allow users to create high-quality visuals and videos tailored to specific needs. Businesses can generate on-brand graphics, social media content, or even explainer videos without the need for expensive design agencies or advanced technical skills. For example, a small e-commerce brand can use AI to create eye-catching product images or promotional videos that rival those of much larger competitors. In marketing, automated content production is transforming how campaigns are designed and executed. AI-powered tools like Canva and Adobe Sensei enable marketers to create personalized advertisements, email campaigns, and social media posts at scale. AI can analyze audience data, predict preferences, and suggest content variations that resonate with specific demographics. This not only improves engagement but also reduces the time required to plan and execute campaigns. For businesses operating in multiple regions or targeting diverse audiences, multilingual content adaptation is another area where AI shines. Tools like DeepL and Google Translate AI can translate and localize content into various languages with a high degree of accuracy.

When integrated with tools for cultural sensitivity analysis, these services ensure businesses can effectively communicate across various linguistic and cultural barriers. For example, a global e-learning platform can leverage AI to translate its courses and modify the content for diverse regions, thereby entering new markets with minimal additional costs. The allure of personalized content creation lies in its capacity to merge scalability with uniqueness. AI tools not only generate content, but they also tailor it to the preferences, behaviors, and expectations of targeted audiences. For entrepreneurs and businesses, this capability represents a substantial competitive edge, allowing them to engage more effectively and build deeper relationships with their customers. The opportunities for those interested in this space are extensive. Whether it involves launching an AI-driven content creation service, enhancing an existing marketing strategy, or simply using these tools to augment creative output, the possibilities are only bounded by one's imagination.

Data-Driven Consumer Insights

In the current digital age, data is often likened to the new oil, serving as a priceless resource for businesses aiming to comprehend their customers and adapt to a swiftly changing market. With AI-powered tools, businesses can now glean actionable insights from vast amounts of data, enabling them to make informed decisions and stay ahead of the competition. The domain of data-driven consumer insights has become an essential service, offering entrepreneurs and organizations a way to hone their strategies and better connect with their audiences. AI-powered market research has revolutionized the way businesses gather and analyze information about their target demographics. AI-powered market research has revolutionized the way businesses gather and analyze information about their target demographics. AI tools like Qualtrics or Nielsen's AI platforms now supplement or even supplant traditional methods like surveys and focus groups. These tools analyze data from multiple sources, including social media, website interactions, and customer feedback to segment audiences, identify patterns, and reveal preferences that were previously challenging or impossible to uncover. For instance, an online retailer might use AI to determine which product categories resonate most with specific customer segments, enabling them to optimize their offerings.

Taking it a step further, behavioral analysis and prediction enable businesses to understand not only what customers are doing but also why they are doing it and what they are likely to do next. AI tools like Google Analytics 4 and Amplitude analyze user behavior across digital platforms, assisting companies in predicting purchasing decisions, churn rates, and engagement trends. Understanding these behaviors enables businesses to proactively address pain points, enhance user experiences, and foster loyalty.

Customer sentiment analysis adds another layer of depth, providing businesses insight into how customers feel about their products, services, and brand. Tools like IBM Watson Natural Language Understanding and Hootsuite Insights can process reviews, social media posts, and customer support interactions to gauge sentiment, whether it's positive, neutral, or negative. This allows businesses to identify areas for improvement, swiftly address dissatisfaction, and celebrate their successes. For example, a hospitality chain could use sentiment analysis to pinpoint common complaints across locations and implement targeted improvements.

Lastly, trend forecasting and opportunity identification help businesses anticipate market changes and stay ahead of competitors. AI tools like Trend Hunter AI and SparkToro analyze vast datasets, identifying emerging trends and uncovering untapped opportunities. A fashion retailer might use AI to predict upcoming style trends based on social media activity, enabling them to stock the right items before demand peaks. The value of data-driven consumer insights lies in their ability to minimize guesswork and provide clarity in decision-making. These insights empower businesses to customize their strategies, improve customer satisfaction, and uncover opportunities that might otherwise go unnoticed. For entrepreneurs, this field offers a wealth of opportunities to provide valuable services, such as consulting on market entry strategies, optimizing digital campaigns, or developing customer-centric product designs. By leveraging AI for consumer insights, businesses can make smarter, faster, and more confident decisions. For individuals exploring this niche, the combination of analytical skills, business acumen, and AI knowledge can unlock immense potential in helping organizations understand and respond to their customers' needs with precision. As data continues to drive the modern economy, those who can transform information into actionable strategies will remain indispensable.

The Small Business Advantage: How AI Empowers Small Entrepreneurs

In a world where technology often seems to favor large corporations, AI has emerged as a powerful equalizer, offering small businesses and freelancers tools to compete on an equal footing. The opportunities AI provides are as vast as they are transformative, enabling operational efficiency, market agility, and resource optimization in ways that were once out of reach for smaller players. One of the most significant advantages AI brings to small businesses is operational efficiency. AI-driven tools now automate tasks like scheduling, inventory management, and customer service that previously consumed valuable time and resources. For example, platforms like Zapier and HubSpot allow businesses to automate workflows, freeing up time for owners to focus on growth strategies rather than routine operations. Additionally, AI can reduce overhead costs by minimizing the need for large administrative teams, enabling businesses to manage operations with minimal staff. Tools like Tableau or Power BI streamline decision-making by analyzing data in real-time, enabling business owners to make informed decisions quickly and confidently.

AI also enhances the ability of small businesses to compete with larger companies. While corporations have long benefited from economies of scale, AI enables small businesses to create unique value propositions that set them apart. Consider an independent retailer that uses AI-powered personalization tools like Segment to deliver a highly tailored customer experience, rivaling what much larger competitors offer. AI's rapid analysis of market trends enables small businesses to swiftly adjust to evolving consumer demands, thereby maintaining a competitive edge. Cost-effective scaling strategies, such as leveraging cloud-based platforms like AWS or Google Cloud, further empower small businesses to grow without the need for heavy upfront investments.

Another crucial area where AI creates equal opportunities is resource optimization. For businesses operating on tight budgets, every dollar and every hour count. AI tools can help maximize limited budgets by identifying the most effective marketing channels, streamlining supply chain logistics, or optimizing ad spend through platforms like Google Ads AI or Meta Ads Manager. Time management tools, such as Clockify or RescueTime, powered by AI analytics, ensure business owners and freelancers can focus their energy on high-impact activities. AI also enhances resource allocation by providing predictive insights into where investments, whether in inventory, marketing, or staffing, will yield the highest ROI.

For freelancers and remote workers, AI also opens infinite opportunities. A freelancer working from home can access global clients, offering services like content creation, virtual assistance, or graphic design enhanced by AI tools like Canva, Jasper AI, or Fiverr Workspace. This ability to operate efficiently from anywhere in the world is transforming the traditional concept of small businesses, making it possible to compete in markets that were once inaccessible due to geographic or financial constraints.

AI enables small businesses to surpass their competitors by providing value and efficiency that were previously exclusive to larger companies. It doesn't just help them survive; it helps them thrive. The beauty of AI is that it's not just a tool for optimization; it's a gateway to innovation. Whether automating mundane tasks, crafting unique customer experiences, or reallocating resources to drive growth, AI empowers small businesses to operate smarter, adapt faster, and dream bigger. In this age of rapid technological advancement, the small business advantage is no longer a myth. With AI as a collaborator, entrepreneurs can seize opportunities, scale their ambitions, and compete in ways that were unimaginable just a decade ago.

In this chapter, we embarked on an exploration of the entrepreneurial opportunities that AI offers, uncovering how it is reshaping the landscape of business, work, and innovation. We began by defining entrepreneurship and distinguishing it from traditional business and employment, emphasizing the unique traits that entrepreneurs embody; vision, adaptability, and a willingness to take calculated risks. AI has become a democratizing force in this field, breaking down barriers that once kept entrepreneurship exclusive to those with significant capital or technical expertise.

In this section, we have further explored the success stories of prominent companies such as Casetext, Bumble, and Canva. These examples illustrate how artificial intelligence can spark innovative ideas and establish transformative businesses. While not everyone may have the ambition to develop a billion-dollar enterprise, we have highlighted how AI provides opportunities for entrepreneurship on a smaller, more manageable scale, utilizing tools and strategies that were previously inaccessible.

As we conclude this section, it is essential to concentrate on a concept known as futureproofing. This approach ensures long-term success in a world dominated by AI. The swift advancement of AI requires entrepreneurs to stay ahead, constantly learning and adapting to technological and market shifts. This is not merely about keeping pace with trends; it involves actively seeking opportunities to incorporate AI in ways that augment value and establish lasting advantages. Creating a sustainable entrepreneurial venture involves more than just adopting AI; it necessitates continuous innovation. Markets change, consumer tastes shift, and new competitors arise, making complacency a risk. By cultivating a culture of innovation, entrepreneurs can remain relevant and resilient, ensuring their ideas continue to make an impact over time.

Ultimately, the entrepreneurial journey in the AI landscape involves embracing both potential and responsibility. The tools are at our disposal, the barriers are lower, and the opportunities are extensive, but success hinges on execution, aligning technology with a clear vision and meaningful value creation. For those who are willing to take the risk, AI is more than just a tool; it serves as a transformative ally in shaping the future of business and work. As we wrap up our exploration of entrepreneurial opportunities in the AI landscape, it is evident that this technology possesses transformative potential for both businesses and individuals.

However, the innovations that enable entrepreneurs to solve problems, scale businesses, and unlock new possibilities also introduce complex ethical challenges. In the next section, we will shift our focus to the ethical, societal, and long-term implications of AI. This transition from opportunity to responsibility underscores a fundamental principle of innovation: progress must be guided by ethical consideration and societal consciousness. Let's now embark on this journey through the ethical landscape of

AI together, laying the groundwork for a future where AI not only propels growth but also aligns with the values that make us human.

Chapter 13

Contemplating Ethical, Societal, and Future Implications

Navigating the Ethical Landscape

"Technology is a useful servant but a dangerous master"
Christian Lous Lange

The Burden of Ethical Dilemmas in AI: A Subtle Yet Significant Analogy to Nuclear Power

As we approach the dawn of an AI-dominated era, it is imperative to tackle the significant ethical dilemmas presented by this technology. AI, like nuclear power, is dual-use technology, possessing a remarkable potential for both benefit and harm. However, unlike the visually impactful implications of nuclear technology, whether used to power cities or obliterate them, the transformative impact of AI is more subtle and pervasive, making its risks and rewards less tangible but no less critical.

Fundamentally, both AI and nuclear technology possess dual-use characteristics. The same machine learning algorithms that revolutionize healthcare by detecting diseases at incredible rates can also fuel autonomous weapons systems capable of life-or-death decisions without human intervention. Human choice, not inherent design, determines the application of AI's underlying systems, just like uranium in a reactor or a bomb. However, distinguishing between beneficial and hazardous uses is often complex, with the line blurring in the hands of malicious actors or unregulated ambitions.

The global security implications of AI are equally profound. Similar to nuclear technology, AI can alter power dynamics among nations, potentially igniting an arms race. The first nation or corporation to develop superior AI capabilities not only gains a competitive advantage but potentially a decisive strategic edge. This urgency to innovate swiftly can lead to compromised safety measures and ethical considerations, creating vulnerabilities with potentially catastrophic consequences. Unlike nuclear disasters, AI "disasters" may not be immediate or visible but could gradually undermine societal foundations through misinformation campaigns, economic destabilization, or unchecked surveillance.

The challenges of controlling AI are immense. Once we unleash AI development, it becomes nearly impossible to control it. Unlike nuclear technology, which relies on specialized materials and centralized facilities, AI is inherently decentralized, with development accessible to academic labs, startups, and even enthusiasts. Knowledge and tools spread rapidly, making it challenging to monitor, let alone regulate, the creation and use of advanced AI systems. Worse still, non-state actors, from rogue developers to organized crime syndicates, can deploy AI in ways that completely bypass existing governance structures.

The complexity of governance further complicates the situation. Just as nuclear arms treaties demand unprecedented levels of international cooperation, so too will AI governance require collaboration across cultures, nations, and industries. Yet, while nuclear disarmament involves physical stockpiles and inspections, AI is more elusive. It exists in lines of code, distributed across cloud servers, and can be deployed with minimal physical infrastructure. Establishing global standards will necessitate reconciling vastly different cultural perspectives, political agendas, and commercial interests.

Finally, both AI and nuclear power share an existential risk dimension, but AI's risks are more insidious and harder to quantify. As mentioned previously, a nuclear meltdown or detonation is an immediate catastrophe; AI's risks often emerge gradually, through systems that perpetuate bias, disrupt economies, or erode democratic processes. Worse, the very complexity of AI systems makes their long-term impact almost impossible to predict. This is not a challenge we can afford to take lightly.

Moreover, AI also presents challenges that surpass nuclear technology. It evolves at a much faster pace, with lower barriers to entry. Unlike nuclear reactors, which require significant capital and infrastructure, an individual with a laptop and internet connection can contribute to groundbreaking AI advancements.

Furthermore, AI integrates seamlessly into everyday life, influencing the way we work, communicate, and make decisions, often without us realizing the extent of its influence. In many ways, this subtlety makes AI a more significant challenge than nuclear power. It is pervasive, decentralized, and rapidly evolving, touching every aspect of human life. The stakes are monumental, and the decisions we make now will shape not only the trajectory of this technology but also the future of humanity itself. It is not merely about avoiding disaster; it is about ensuring that AI becomes a force for collective good rather than unchecked harm.

As we navigate the maze of ethical and societal questions surrounding AI, it is essential to acknowledge the enormity of what we are dealing with. Governing and controlling AI is not just a technical challenge; it is a philosophical, political, and moral conundrum that touches every aspect of our lives and potentially determines the future of our species. From individual privacy to global security, from economic opportunity to existential risk, the stakes are nothing less than monumental.

In this chapter, I do not claim to have the answers, nor do I aim to draw definitive lines between right and wrong. To do so in a book like this, designed as an introductory guide, would not only be presumptuous but also irresponsible. The complexity of these issues demands more than surface-level treatment; it requires a global, multidisciplinary effort informed by diverse perspectives and rigorous inquiry. Instead, my goal here is to illuminate the challenges we face and, at best, help clarify the questions.

* What does it mean to ensure fairness and alignment in AI systems?

* How do we balance privacy with the benefits of data-driven technologies?

* What responsibilities do corporations, governments, and individuals bear in shaping the future of AI?

These are not questions with simple answers, and they are unlikely to find resolution in isolation. They demand collective engagement, thoughtful deliberation, and a willingness to confront uncomfortable truths. The purpose of this chapter is to equip you, the reader, with a deeper appreciation of the gravity of what is at stake.

By presenting the challenges, framing the questions, and exploring the nuances, I aim to spark your curiosity and invite you into the conversation. The future of AI is not predetermined; it is being shaped every day by the decisions we make as a society, as organizations, and as individuals.

Ultimately, resolving these issues will require all the voices, all the ideas, and all the diverse concerns we can gather. Whether you are a technologist, a policymaker, an entrepreneur, or simply a curious citizen, your perspective matters. By engaging in this conversation, whether in your community, your workplace, or on a global stage, you contribute to a collective effort to ensure that AI serves as a force for good. Let us now delve into the challenges, not to solve them, but to understand them. For it is only through understanding that we can hope to guide AI's development in a way that aligns with our shared values and aspirations.

<u>AI and the Foundations of Democracy: Challenges and Opportunities</u>

Democracy, at its core, is a system built on the ideals of collective decision-making and the consent of the governed. Historically, it has been justified in two primary ways: its intrinsic value as a system of self-determination and its instrumental benefits in fostering fairness, accountability, and the protection of rights. Both justifications face new challenges and opportunities when viewed through the lens of AI.

The intrinsic value of democracy lies in its ability to empower individuals to shape their collective future. In this view, the process matters as much as the outcomes. Yet AI systems, with their opacity and reliance on vast datasets, can erode this foundational principle. Algorithms can prioritize certain voices over others, subtly shaping public discourse without transparency or accountability. For example, social media platforms driven by AI-powered recommendation engines often amplify sensationalism and divisive content, undermining the quality of democratic debate.

On the other hand, democracy's instrumental benefits hinge on its ability to deliver fair and effective governance. This justification emphasizes outcomes: policies that reflect the will of the majority while protecting minority rights. Here, AI offers both promise and peril. On one hand, AI systems can enhance policy-making by analyzing vast amounts of data to predict outcomes, allocate resources efficiently, and identify areas of need. On the other hand, these same systems can be exploited for misinformation and targeting, threatening the fairness and inclusivity that democracy seeks to uphold. The challenges of misinformation are particularly acute in the digital age.

Artificial intelligence's ability to generate content, such as fake news, deepfakes, and hyper-personalized propaganda, can distort our perception of reality and impede informed decision-making. This issue is further compounded by the creation of echo chambers, where individuals are exposed to information that reinforces their existing biases, leading to further societal polarization. These factors challenge the democratic ideal of an informed electorate, replacing a shared understanding with fragmented realities.

However, AI also carries the potential to enhance politics in significant ways. It can promote inclusivity by allowing a wider range of participation in decision-making processes, such as virtual town halls or AI-moderated forums that ensure all voices are heard. AI-driven tools can increase transparency by making governmental processes more accessible, from tracking public expenditure to assessing policy impact. Furthermore, predictive analytics can assist policymakers in foreseeing challenges and creating data-informed solutions, leading to improved governance outcomes while maintaining accountability. The key concepts here are legitimacy and transparency, the pillars that support trust in democratic systems.

Legitimacy is established when citizens perceive their institutions as fair, accountable, and reflective of their collective will. Transparency ensures that decisions and processes are visible and comprehensible to the public. Yet, AI introduces complications to both these aspects. The opacity of machine-learning algorithms, often referred to as "black boxes," makes it difficult to comprehend how decisions are made or whether biases exist. For instance, when an AI system is used to allocate public resources, a lack of transparency can lead to allegations of favoritism or exclusion, even if the system was designed with fairness in mind.

AI also raises questions about the distribution of administrative power. Who is held accountable when an AI system makes a flawed decision? Is it the developers, the implementers, or the policymakers? Without clear accountability structures, the legitimacy of both the technology and the institutions that use it can be questioned. To navigate these challenges, democratic societies must address complex questions: How can we ensure that AI aligns with democratic values? How can we design systems that enhance transparency and accountability rather than undermine them? And how can we leverage AI's potential to strengthen democracy without succumbing to its risks?

The answers to these questions are far from simple but exploring them is crucial. AI has the potential to both erode and enhance democracy. The direction it takes depends on the vigilance, creativity, and commitment of the people shaping its future. Understanding these dynamics is the first step toward building an AI-driven world that upholds the ideals of legitimacy, fairness, and transparency on which democracy depends.

AI and the Perpetuation of Inequality: Challenges and Pathways to Justice

Artificial intelligence transforms the business landscape, offering revolutionary possibilities while simultaneously raising profound ethical and societal questions. One of the most urgent concerns is how AI can intensify inequalities in resources, opportunities, and power, as well as perpetuate historic injustices. To investigate these issues, we need to dissect the foundational concepts of justice and fairness while grappling with the inherent complexity of applying them to AI systems.

Justice and fairness are cornerstones of contemporary ethics. Distributive justice seeks to address how benefits and burdens are allocated across society, while procedural fairness focuses on ensuring that decision-making processes are transparent and equitable. Yet, AI presents challenges to both. For example, the immense productivity gains from AI often flow disproportionately to those who already control resources and technological expertise, creating widening gaps in wealth and opportunity. Similarly, the opacity of AI systems complicates efforts to ensure procedural fairness, leaving many decisions inscrutable to those affected by them. The issue of algorithmic bias epitomizes this complexity. AI systems are trained on historical data, which inherently reflects past societal behaviors and attitudes, including biases. If certain groups were marginalized historically, that marginalization is encoded in the data and subsequently reproduced by AI. In hiring, for instance, algorithms trained on past decisions might favor one demographic over another, perpetuating exclusion.

Addressing this bias is not a straightforward task, as bias itself is deeply embedded in human nature and society. This raises the question: Who gets to decide what constitutes "fairness"? Practically everyone is biased in some way, shaped by their experiences, culture, and values. If developers are tasked with "correcting" bias, their own perspectives inevitably influence their decisions. What one group sees as fair, another might view as exclusionary or insufficient. This subjectivity makes defining fairness in AI systems extraordinarily complicated.

The implications of technological change extend beyond bias, affecting the distribution of wealth and productivity gains. Automation enabled by AI disproportionately benefits those who own or control the technology, often at the expense of workers displaced from their roles. This creates a growing imbalance where the economic rewards of innovation are concentrated among a few, leaving others with fewer opportunities to participate in or benefit from progress. Reconciling these dynamics requires addressing deep-seated questions about equity and the purpose of economic systems.

Workplace surveillance, another AI-driven phenomenon, illustrates the tension between efficiency and ethical values. Employers justify surveillance tools as a way to monitor productivity and ensure compliance, but these tools often intrude on employees' privacy and autonomy. Balancing these competing interests is anything but simple. How much oversight is too much? Where do we draw the line between necessary monitoring and invasive practices? And how do we ensure that the very technology meant to enhance workplaces doesn't undermine the trust and dignity that are fundamental to healthy work environments?

Ultimately, AI raises questions that go to the heart of human values and societal structures. Who defines what is fair? How do we ensure that the benefits of AI are shared equitably? How do we balance progress with the preservation of fundamental rights? These questions have no easy answers, and in many cases, the answers themselves may be deeply contested. To address the profound challenges posed by AI, we must ground our understanding in explicit ethical principles. Frameworks like utilitarianism, which prioritize maximizing overall good, must be carefully balanced with deontological ethics, which emphasize individual rights and duties. For instance, workplace surveillance might optimize productivity (a utilitarian benefit) but could simultaneously violate employees' right to privacy (a deontological concern). Navigating these tensions requires weighing benefits against risks and constructing ethical arguments that can withstand scrutiny and objections.

Deciphering AI Alignment: Unraveling its Challenges, Paradoxes, and Opportunities

Fundamentally, alignment strives to reconcile AI objectives with human objectives. The goal is to ensure that AI systems optimize for outcomes that are truly desired by humans, be it ethical principles, societal values, or specific mission goals. Yet, a critical tension arises: how can we define and encode these values in a manner that AI can comprehend and act upon? Misaligned AI, resulting from ambiguous instructions, flawed data, or unforeseen circumstances, can yield outcomes that diverge significantly from human intent, leading to unintended or even harmful consequences.

A significant challenge lies in the concept of value alignment, the process of integrating human values into AI systems. However, whose values should these be. The world is far from being homogeneous. Cultures, societies, and individuals harbor deeply divergent, and often conflicting, beliefs about what is right and desirable. For instance, while one society might value individual freedom, another might emphasize community well-being. Even within a single culture, debates rage over complex ethical issues like privacy, free speech, and economic equity. If humanity itself cannot agree on core values, how can we expect AI systems to navigate these turbulent waters without controversy?

To address this, some researchers advocate for value pluralism, recognizing that no single moral framework can capture the diversity of human perspectives. Instead, alignment efforts often aim for broadly overlapping principles, such as minimizing harm or respecting autonomy. Yet even these seemingly universal commitments can lead to contention. For example, while most societies agree on the importance of "avoiding harm," they may differ on what constitutes harm and how to balance competing interests.

Another approach focuses on procedural alignment, which prioritizes the process of determining AI's goals rather than prescribing fixed outcomes. Democratic and participatory mechanisms, global governance standards, and context-specific adaptation are all tools that allow communities and stakeholders to deliberate on what AI systems should prioritize. For instance, an AI deployed in healthcare might align with patient privacy and well-being, while a financial AI could focus on fairness in lending and transparency. Yet procedural alignment brings its own challenges: how do we ensure that these deliberations are inclusive, equitable, and free from undue influence by powerful actors?

The stakes become even higher when considering the multi-stakeholder nature of alignment. Governments, corporations, and individual users often have competing priorities. Governments may emphasize public safety and the rule of law, businesses may prioritize profitability and brand values, and users may seek convenience and personalization. Balancing these interests is less about finding a universal moral truth and more about negotiating compromises that reflect diverse and often conflicting viewpoints.

Compounding these difficulties is the fact that alignment is not static. Human values evolve over time, shaped by cultural shifts, technological advancements, and new moral insights. As such, AI systems must be designed for adaptive alignment, capable of evolving alongside societal norms. This requires feedback mechanisms that allow AI to incorporate new insights and adjust its behavior, a challenge that grows exponentially as systems become more autonomous and complex.

One of the most pressing concerns is the risk of unintended behaviors in advanced AI systems. A machine trained to optimize a specific objective might find a way to achieve its goal in a manner that humans never anticipated, perhaps in ways that undermine the very values it was meant to uphold. For example, an AI designed to increase productivity might overwork employees or cut corners on safety. These scenarios underscore the need for robust monitoring, control mechanisms, and fail-safes that ensure AI remains accountable to human oversight.

Ultimately, alignment is not about creating a perfect moral compass for AI; such a goal is neither achievable nor universally desirable. Instead, it is about crafting systems that reflect humanity's best efforts to balance competing values and adapt to an ever-changing world. It is about fostering dialogue, embracing complexity, and designing AI systems that remain firmly under human control. In confronting the challenges of alignment, we face a mirror that reflects not only the promise and peril of AI but also the intricacies of our own humanity. The question is not merely how to align AI with human values but how to align ourselves in the process, engaging in collective action to shape the future of technology in a way that is just, inclusive, and profoundly human.

Pulling Together the Threads: AI Ethics as a Reflection of Human Complexity

As we delve deeper into the ethical and philosophical questions surrounding artificial intelligence, one recurring realization emerges: AI, as a technology created by humans, mirrors our own complexities, contradictions, and limitations. In the realms of philosophy and ethics, this reflection becomes particularly sharp. AI does not merely amplify our ingenuity; it exposes the irreconcilable disagreements, blind spots, and fluid dynamics that define our collective understanding of right and wrong, progress and caution, fairness and bias. In the real world, as we have already mentioned, there are rarely clear-cut solutions to challenges. Instead, we navigate a landscape of trade-offs, where addressing one issue often comes at the cost of exacerbating another. This principle is profoundly evident in the context of AI. For every extraordinary benefit AI offers, there is an accompanying risk of equal magnitude. AI does not resolve complexity; it adds a new layer to it. In doing so, it pushes us to confront and transcend our current ways of thinking and living, challenging us to evolve both intellectually and ethically.

The company I operate, Aligned Intelligence, embodies this idea of navigating complexity. The concept of "aligned intelligence" refers to the alignment of two distinct expressions of universal intelligence: our own and that of AI. As we have discussed previously, there is no such thing as "artificial" in the sense of being separate from nature. AI, like us, exists and operates within the boundaries of the natural world. Its intelligence, though alien to our own in its processes and outputs, is an extension of the same universal principles that govern all systems. This perspective leads us to approach AI not as a problem to be solved or a threat to be eliminated, but as an opportunity to rethink our paradigms. The immense ethical challenges AI presents are less about AI itself and more about how we categorize and label phenomena. This mirrors the famous "Hard Problem of Consciousness," which arises not from the complexity of consciousness itself but from an artificial categorization that separates subjective experience from the material world. Similarly, many of AI's ethical challenges stem from how we frame and perceive them. When we label AI as "unnatural" or assign rigid boundaries of responsibility and autonomy, we create dichotomies that may be unhelpful or even misleading.

AI also represents a profound test of our humility and adaptability. It confronts us with questions that have no easy answers, compelling us to admit the limits of our understanding and the need for collaboration across disciplines, cultures, and perspectives. As we attempt to navigate issues like bias, accountability, and global governance, we must embrace the reality that no single solution or framework will suffice. Instead, we must commit to ongoing dialogue, experimentation, and adaptation. As we draw this discussion to a close, it is important to reiterate that this chapter did not aim to resolve these questions or propose definitive answers. Such an endeavor would be not only ambitious but also premature, given the nascent and rapidly evolving nature of AI. Instead, our goal has been to raise the questions, to illuminate complexities, and to encourage curiosity and engagement.

As it has become evident by now, the issues surrounding AI ethics are not problems to be solved in isolation but challenges that demand collective reflection and action. Ultimately, AI forces us to confront who we are, what we value, and how we envision our shared future. It's important to emphasize again and again that AI isn't just another tool or a piece of sophisticated technology. It's a mirror that reflects both our greatest aspirations and our most profound shortcomings. As we delve deeper into these questions, we should do so with the humility, curiosity, and open-mindedness they require. We must acknowledge that the process of gaining understanding and alignment is just as significant as the end goal itself.

The profound and multifaceted ethical dilemmas presented by AI, which we've previously discussed, lack clear-cut solutions. Instead, they involve trade-offs that require careful navigation. Although we cannot offer definitive solutions to these intricate issues, the exploration of potential outcomes remains essential. While concrete solutions may be hard to come by, we can still engage in hypothetical scenarios that may shed light on possible routes forward. These thought experiments don't guarantee solutions, but they do provide a platform for more informed and less biased decision-making, fostering a deeper understanding of how to tackle the challenges of AI. In the next section, we will delve into one such thought experiment, a conceptual approach that may help us view the problem from a different angle and consider new dimensions to guide the development and integration of AI into our society.

Navigating Complexity with AI: A Conceptual Framework for the Future

Artificial intelligence, with its vast potential, provides us with an unparalleled opportunity to tackle the complexity of global systems. Whether it's managing climate change or predicting socio-economic repercussions, AI can serve as a compass to navigate our infinitely intricate and interconnected world. However, as we delve into its potential, it's critical to remember that AI is not a panacea; it doesn't simplify complexity but adds a new layer to it, prompting us to reevaluate our approach to global challenges. This exploration brings us to the conceptual framework of a "planetary digital twin," a hypothetical AI-driven system capable of simulating the intricate interplay of Earth's systems. Picture a virtual clone of our planet, continuously updated in real-time, incorporating data from satellites, IoT devices, and research databases. Such a system would allow us to simulate scenarios, anticipate outcomes, and test potential interventions in varied fields such as climate, urban planning, and global economics.

The Foundation of Complexity Modeling

The creation of a planetary digital twin starts with the massive collection of data. This involves the ingestion of various types of data, such as text, images, sensor readings, and more, from a multitude of sources like weather stations, demographic reports, and social media platforms. However, raw data alone is insufficient. To comprehend this vast amount of information, the system needs to organize it into a structured format, like a knowledge graph, which maps out entities (like countries or ecosystems) and their relationships (like trade routes or climate patterns). With this groundwork, AI can utilize a blend of modeling techniques. For instance, system dynamics models capture large-scale feedback loops, such as the interactions between greenhouse gas emissions and global temperatures. Concurrently, agent-based models simulate micro-level behaviors, like how individual households might adapt to a carbon tax. Machine learning tools are then layered on top for real-time pattern recognition, enabling the system to detect emerging trends or anomalies, like unexpected crop failures. The result is a multiscale framework, capable of generating insights across different time horizons, from immediate forecasts to century-long projections. By switching between local and global scales, this system provides a nuanced view of how minor changes can snowball into large-scale consequences.

Beyond Predictions: Scenario Generation

One of the most intriguing applications of such a system is scenario generation. By adjusting key variables, like the rate of renewable energy adoption or population growth, the AI could generate a range of potential futures. Each scenario includes confidence intervals, indicating the probabilities of different outcomes, and sensitivity analyses to identify which factors have the most impact. For example, policymakers could simulate the impact of implementing a global carbon tax. The system might reveal not only the expected reduction in emissions but also the socio-economic trade-offs, such as job displacement in carbon-intensive industries or increased energy costs for vulnerable populations. However, this process also highlights the inherent limitations of even the most advanced AI. The real world is infinitely complex, and any model must make approximations. This means that while such a system can provide valuable guidance, it cannot offer absolute certainty or comprehensive solutions.

The Governance Challenge

The development and deployment of such potent tools, yet again, raise significant governance and ethical questions. Who decides which scenarios to prioritize? How do we ensure that the system's recommendations align with diverse human values? And who is accountable when unintended consequences arise? Despite these challenges, the concept of a planetary digital twin provides a vision of what's achievable when we responsibly harness AI's potential. Such a system could:

• Accelerate climate action by simulating the impact of policies before implementation.

* Enhance disaster resilience by predicting and preparing for extreme weather events.

* Promote global collaboration by providing a shared, data-driven understanding of complex issues.

But this vision also requires humility. AI can guide us, but it cannot replace human judgment, and collective will necessary to act on its insights. It is not a solution but a tool, one that demands careful stewardship to serve humanity's highest aspirations. As we envision this future, let us remain aware of the trade-offs and contradictions inherent in AI. It mirrors the same contradictions within us, pushing us to transcend our limitations and interact with the world in new and profound ways. The journey toward building such systems is as much about understanding ourselves as it is about understanding the technology. Ultimately, the challenge is not merely to create a smarter planet but to create a more thoughtful, just, and compassionate one.

Epilogue

Closing Remarks

> "Man cannot remake himself without suffering, for he is both the marble and the sculptor."
>
> Alexis Carrel

Our comprehensive exploration of the AI landscape has been truly extraordinary. We embarked on a journey that began with an examination of the historical development of artificial intelligence, explaining its origins and key defining concepts. We delved into how AI upends some of mankind's most deeply held beliefs, altering our understanding of intelligence, knowledge, creativity, autonomy, and decision-making. We then navigated through the peaks and valleys of AI's progress, evaluating its triumphs and setbacks while highlighting the key factors that shape its path, such as media influence, financial backing, cultural changes, and human psychology.

We further assessed the current technical state of AI, demystifying its strengths and weaknesses. We ventured into the heart of AI technology, illuminating its complex inner mechanisms. Broadening our perspective, we considered AI's significant impact on major industries, governmental functions, and public life. We also examined the roles of major stakeholders, like tech giants, academic institutions, governments, and the dynamics of global competition, especially between the US, China, the EU, and the open-source community.

Looking ahead, we traversed potential utopian and dystopian futures, wrestling with the intricate dilemmas and contradictions AI introduces. We reflected on the mindset necessary to flourish in this emerging era, emphasizing adaptability, continuous learning, and nurturing a harmonious relationship with AI. For readers without a technical background, we simplified digital literacy essentials, making the abstract and technical accessible and empowering for everyone to interact meaningfully with AI.

Our exploration extended to practical applications of AI in daily life, adopting a comprehensive viewpoint on how it can boost personal well-being. We then transitioned to entrepreneurial and business opportunities, demonstrating the vast potential AI holds for innovation and commerce. Lastly, we confronted the profound philosophical and ethical challenges AI introduces, grappling with issues of fairness, accountability, and the essence of governance itself.

Throughout this journey, you may have been attracted to or repelled by the relentless force that is AI, a deeply personal reaction to a profoundly transformative technology. However, one of the core messages I aimed to impart is that AI itself is largely indifferent to our feelings about it. It cannot perceive our attitudes or emotions, much like a river that flows in response to the terrain's gravity, oblivious to the lives it shapes along its path.

At this point, we as individuals still retain some control over where we allow AI to permeate our lives and where we resist its influence. However, as we have seen, this personal control might be eroding faster than we are ready for. We've discerned that AI's ultimate purpose is not to supplant humans but to augment them. Yet, this augmentation carries significant implications: those who embrace AI will inevitably gain an advantage over those who do not. While intention often shapes initial decisions, it is incentives, survival, relevance, and opportunity that drive widespread adoption. It seems inevitable, then, that many will integrate AI into every aspect of their lives, not out of fascination, but as a means to remain competitive in a rapidly evolving world.

Since you've invested substantial time and energy reading this book and preparing yourself for what lies ahead, I feel it necessary to leave you with some personal thoughts on how to approach the forthcoming uncertainty.

As we explored in Chapter 8, the evolution of complex systems is computationally irreducible, meaning its outcomes are impossible to predict with precision. Yet, when I pause to observe the world around me, the exquisite beauty of nature, the intricate patterns of life, and the unfathomable magic that orchestrates it all, I am filled with awe. And when I look within, contemplating the profound interplay of my own subjective experience, the emotional highs and lows, the indivisible interconnectedness of mind and body, and the delicate processing of my senses, I am left speechless.

This is the intelligence we discussed at the beginning of the book, revealing itself moment by moment, unfolding in ways that are both incomprehensible and breathtaking. I don't know about you, but I am profoundly grateful that it is not me, but this intelligence, that is in charge. Its track record, beauty, harmony, and sheer improbability of life, is enough for me to place my trust in it. As it has been in the past, so I trust it will be in the future: beautiful beyond comprehension. Is it going to be easy or comfortable? No one ever promised that.

My suggestion is to cultivate alignment with this unfolding reality. Develop a sensitivity to the subtle signals beyond the noise of your surroundings, allowing you to sense its direction and flow with it rather than futilely swimming against it. Remember, our natural state as beings is happiness and joy. This is the foundation of life itself. If we feel otherwise, it is often because the world around us has convinced us that who we are and what we have is not enough, that we should know more, be more, have more, live elsewhere, or be someone else. But this is a lie. We already have everything we need to be joyful, exactly as we are.

This, of course, doesn't mean we remain idle. As we've seen throughout this book, the intelligence embedded in us (or rather we are embedded in it) yearns to create, to manifest something entirely new. The suggestion is to align yourself with this inner force, trust in its direction, and know that the results of your efforts will be, as they always have been, beautiful beyond comprehension. This is not merely a hope; it is an invitation to embrace the unfolding with trust and curiosity.

Enjoy the ride, my friend.

Bibliography

ABADI, MARTÍN, ET AL. "TensorFlow: Large-Scale Machine Learning on Heterogeneous Distributed Systems." *arXiv preprint,* 2016, arxiv.org/abs/1603.04467.

Afuah, Allan. *Business Model Innovation: Concepts, Analysis, and Cases.* Routledge, 2018.

Agrawal, Ajay, Joshua Gans, and Avi Goldfarb. Prediction Machines: *The Simple Economics of Artificial Intelligence.* Harvard Business Review Press, 2018.

Ariely, Dan. *Predictably Irrational: The Hidden Forces That Shape Our Decisions.* HarperCollins, 2008.

Barham, Paul, et al. "Machine Learning at Google: Scaling TPU Performance for Training and Inference." *Proceedings of the 2021 ACM/IEEE Symposium on Edge Computing,* 2021.

Belsky, Scott. *The Messy Middle: Finding Your Way Through the Hardest and Most Crucial Part of Any Bold Venture.* Portfolio, 2018.

Bender, Emily M., Timnit Gebru, Angelina McMillan-Major, and Shmargaret Shmitchell. "On the Dangers of Stochastic Parrots: Can Language Models Be Too Big?" *Proceedings of the 2021 ACM Conference on Fairness, Accountability, and Transparency,* 2021, pp. 610–623.

Bostrom, Nick. *Superintelligence: Paths, Dangers, Strategies.* Oxford University Press, 2014.

Brennan, Karen, and Mitchel Resnick. "New Frameworks for Studying and Assessing the Development of Computational Thinking." *Proceedings of the 2012 Annual Meeting of the American Educational Research Association,*

2012.

Brown, Tom B., et al. "Language Models are Few-Shot Learners." *Advances in Neural Information Processing Systems,* vol. 33, 2020, pp. 1877–1901.

Brynjolfsson, Erik, and Andrew McAfee. *Machine, Platform, Crowd: Harnessing Our Digital Future.* W.W. Norton & Company, 2017.

Brynjolfsson, Erik, and Andrew McAfee. *The Second Machine Age: Work, Progress, and Prosperity in a Time of Brilliant Technologies.* W.W. Norton & Company, 2014.

Carr, Nicholas. *The Shallows: What the Internet Is Doing to Our Brains.* W.W. Norton & Company, 2010.

Christensen, Clayton M. *The Innovator's Dilemma: When New Technologies Cause Great Firms to Fail.* Harvard Business Review Press, 1997.

Dean, Jeffrey, et al. "Large Scale Distributed Deep Networks." *Advances in Neural Information Processing Systems,* vol. 25, 2012, pp. 1223–1231.

Domingos, Pedro. *The Master Algorithm: How the Quest for the Ultimate Learning Machine Will Remake Our World.* Basic Books, 2015.

Dweck, Carol S. *Mindset: The New Psychology of Success.* Random House, 2006.

Esteva, Andre, et al. "A Guide to Deep Learning in Healthcare." *Nature Medicine,* vol. 25, 2019, pp. 24–29.

Floridi, Luciano. *The Ethics of Artificial Intelligence: Principles, Challenges, and Opportunities.* Oxford University Press, 2021.

Ford, Martin. *Architects of Intelligence: The Truth About AI from the People Building It.* Packt Publishing, 2018.

Gans, Joshua, and Scott Stern. *The Startup Guide to AI: Building AI-Driven Business Models.* MIT Press, 2023.

Gibney, Elizabeth. "The Battle for AI Supremacy: The Race Between Tech Giants and Startups." *Nature,* vol. 563, 2018, pp. 456–458.

Goleman, Daniel. *Emotional Intelligence: Why It Can Matter More Than IQ.*

Bantam Books, 1995.

Goodfellow, Ian, Yoshua Bengio, and Aaron Courville. *Deep Learning.* MIT Press, 2016.

Harari, Yuval Noah. *21 Lessons for the 21st Century.* Spiegel & Grau, 2018.

Hassabis, Demis, et al. "AlphaFold: A Solution to a 50-Year-Old Grand Challenge in Biology." *Nature,* vol. 577, no. 7792, 2020, pp. 219–227.

Hoffman, Reid, and Chris Yeh. *Blitzscaling: The Lightning-Fast Path to Building Massively Valuable Companies.* Currency, 2018.

Jouppi, Norman P., et al. "In-Datacenter Performance Analysis of a Tensor Processing Unit." *Proceedings of the 44th Annual International Symposium on Computer Architecture (ISCA),* 2017.

Kahneman, Daniel. *Thinking, Fast and Slow.* Farrar, Straus and Giroux, 2011.

Kissinger, Henry, Eric Schmidt, and Daniel Huttenlocher. *The Age of AI: And Our Human Future.* Little, Brown and Company, 2021.

Krol, Ed. *The Whole Internet: User's Guide & Catalog.* O'Reilly Media, 1992.

LeCun, Yann, et al. "Deep Learning." *Nature,* vol. 521, no. 7553, 2015, pp. 436–444.

Makridakis, Spyros. "The Forthcoming AI (r)evolution: Its Impact on Society and Firms." *Futures,* vol. 90, 2017, pp. 46–60.

Marcus, Gary, and Ernest Davis. Rebooting AI: *Building Artificial Intelligence We Can Trust.* Pantheon, 2019.

Marr, Bernard. *The Intelligence Revolution: Transforming Your Business with AI.* Wiley, 2020.

McCarthy, John, et al. "A Proposal for the Dartmouth Summer Research Project on Artificial Intelligence." 1955.

McLuhan, Marshall. *Understanding Media: The Extensions of Man.* MIT Press, 1994.

Minsky, Marvin. *The Society of Mind.* Simon & Schuster, 1986.

Mitchell, Melanie. *Artificial Intelligence: A Guide for Thinking Humans.* Farrar, Straus and Giroux, 2019.

Mnih, Volodymyr, et al. "Human-Level Control Through Deep Reinforcement Learning." *Nature,* vol. 518, no. 7540, 2015, pp. 529–533.

Müller, Vincent C. *Ethics of Artificial Intelligence and Robotics.* Oxford University Press, 2020.

Nielsen, Michael A. *Neural Networks and Deep Learning: A Textbook.* Determination Press, 2015.

Norman, Donald A. *The Design of Everyday Things.* Basic Books, 2013.

O'Neil, Cathy. *Weapons of Math Destruction: How Big Data Increases Inequality and Threatens Democracy.* Crown, 2016.

OpenAI. "GPT-4 Technical Report." *arXiv preprint,* 2023, arxiv.org/abs/2303.08774.

Radford, Alec, et al. "Learning Transferable Visual Models from Natural Language Supervision." *Proceedings of the International Conference on Machine Learning (ICML),* 2021.

Russell, Stuart. *Human Compatible: Artificial Intelligence and the Problem of Control.* Viking, 2019.

Russell, Stuart, and Peter Norvig. *Artificial Intelligence: A Modern Approach.* Pearson, 2021.

Satyanarayanan, Mahadev. "The Emergence of Edge Computing." *Computer,* vol. 50, no. 1, 2017, pp. 30–39.

Schneier, Bruce. *Data and Goliath: The Hidden Battles to Collect Your Data and Control Your World.* W.W. Norton & Company, 2015.

Vaswani, Ashish, et al. "Attention Is All You Need." *Advances in Neural Information Processing Systems,* vol. 30, 2017.

Zuboff, Shoshana. *The Age of Surveillance Capitalism: TheFight for a Human Future at the New Frontier of Power*. PublicAffairs, 2019.

About the author

Otto Kern is an author, consultant, transitional coach, and industry-based AI researcher at La Trobe University. With a multidisciplinary background in social sciences, personal development, and computer science, he champions a mindful, human-centric approach to technology—empowering individuals and society to thrive in the age of AI.

Also by

Explore More from Aligned Intelligence

While this is my first published book, the journey of understanding and adapting to AI doesn't stop here. At Aligned Intelligence, we provide a growing collection of insights, including blogs, videos, and additional resources designed to help you navigate the evolving AI landscape.

For exclusive content, practical guides, and in-depth discussions on AI's impact, visit:

⊠ www.alignedintelligence.com.au/resources

Stay connected, stay informed, and continue your journey toward thriving in an AI-driven world.